Law, Normative Pluralism, and Post-Disaster Recovery

Vivencio O. Ballano

Law, Normative Pluralism, and Post-Disaster Recovery

Evaluating the Post-Disaster Relocation and Housing Project of Typhoon Ketsana Victims in the Philippines

Springer

Vivencio O. Ballano
Department of Sociology
College of Social Science
and Development
Polytechnic University of the Philippines (PUP)
Manila, Philippines

ISBN 978-981-10-5073-2 ISBN 978-981-10-5074-9 (eBook)
DOI 10.1007/978-981-10-5074-9

Library of Congress Control Number: 2017943092

Printed on acid-free paper

This Springer imprint is published by Springer Nature
The registered company is Springer Nature Singapore Pte Ltd.
The registered company address is: 152 Beach Road, #21-01/04 Gateway East, Singapore 189721, Singapore

Acknowledgements

This monograph on law, normative pluralism and post-disaster recovery in the Philippines is a product of a long and painstaking fieldwork, sociological research, and writing. This journey would not have been possible without the generous help of the following people:

To Dr. Ricardo G. Abad, Chairperson of the Sociology-Anthropology Department at the Ateneo de Manila University, my wise advisor and friend throughout my sociological training, for your continuous support and inspiration.

To Dr. Emma Porio, my former sociology professor at the Ateneo de Manila university, for giving some advice on disaster management research that helped me in accomplishing this book.

To Mr. Roel Santos and his staff at the Municipal Disaster Risk Reduction Management Office (MDRRMO) in Rodriguez, Rizal, for the valuable information you shared for this book.

To the Municipal Chairperson and staff of the National Housing Authority (NHA), Rodriguez Office, for sharing some information on the relocation and housing of disaster victims in Southville 8 Housing Project in Barangay San Isidro, Rodriguez, Rizal.

To the Chair and staff of the Housing and People's Development Office (HPDO) of Rodriguez, Rizal, for some data and information on the relocation of disaster victims in the municipality.

To the Administrator of Barangay San Isidro, Rodriguez, Rizal, for sharing some demographic information as well as the social problems encountered by disaster victims in Southville 8 Housing Project.

To my key informants in Southville 8 Housing Project, Phases 8A, 8B, 8C, for sharing your stories and struggles in dealing with the government's inadequate housing and resettlement.

To Prof. Camille Ocampo, Chair of the Sociology Department, and to my co-faculty at Polytechnic University of the Philippines (PUP), Manila, for their support and encouragement.

To Ms. Jayanthie Krisnan, my supportive Publishing Editor at Springer Nature Singapore, and her team, for their enthusiasm and patience in publishing this book.

To my wife, Emily, and my children, Joanne Faye and Johann Karl, for their loving support and inspiration.

Finally, to our Lord Jesus and Mama Mary, for the blessings and guidance in my apostolate of writing—my way of serving the Church and society.

Contents

Abbreviations

BBB	Build Back Better
COA	Commission on Audit
DENR	Department of Environment and Natural Resources
HDMF	Home Development Mutual Fund
HGC	Home Guaranty Corporation
HPDO	Housing and People's Development Office
HUDCC	Housing and Urban Development Coordinating Council
HULRB	Housing and Land Use Regulatory Board
IFRC	International Federation of Red Cross
JV	Joint Venture
LGU	Local Government Unit
LRA	Land Registration Authority
NDRRMC	National Disaster Risk Reduction and Management Council
NEDA	National Economic and Development Authority
NGO	Non-Government Organization
NHA	National Housing Authority
NHMF	National Home Mortgage Finance
PD	Presidential Decree
PDR	Post-Disaster Recovery
PDRRMA	Philippine Disaster Risk Reduction Management Act
PHIVOLCS	Philippines Institute of Volcanology and Seismology
PNP	Philippine National Police
PPP	Public Private Partnership
SHFC	Social Rodriguez Housing Finance Corporation
SRHP	Southville Rodriguez Housing Project
UDHA	Urban Development and Housing Act
UN	United Nations

Chapter 1
Introduction

Abstract This chapter provided an overview of the objective and structure of the book as well as of the one-site case study on which it is based. In particular, it proposed a framework on how the multiplicity of laws and intermediary informal normative orders in the Philippine housing and relocation system has created negative unintended effects to the government's Post-Disaster Recovery (PDR) program under the 2010 Philippine Disaster Risk Reduction Management Act (PDRRMA). Applying the sociological and normative legal pluralist perspectives and the one-site case study method, it outlined broadly the theoretical and methodological framework for evaluating the adequacy of the housing and resettlement project of the Tropical Storm Ketsana victims in the government-owned Southville Rodriguez Housing Project (SRHP), one of the many Southville housing projects established by the government in various parts of the country. Lastly, it showed a roadmap of the book, explaining briefly the content of each chapter and how the chapters are interrelated to achieve the book's overall objectives.

Keywords Tropical Storm Ketsana · Project management · Disaster management · Legal pluralism · Normative pluralism · Post-disaster recovery

1.1 Introduction

Can a plurality of normative orders, both formal and informal, legal and non-legal, in the implementation of a government's post-disaster recovery (PDR) project under the state's official disaster-management law become a deciding factor in the adequacy and the success of a post-disaster housing settlement for typhoon victims?

Exploring the role of the plurality of official laws and informal normative structures behind government disaster projects has been neglected in post-disaster research and management. Evaluating the adequacy and the success rate of the government's PDRs are usually seen from conventional project-management perspectives and rarely from the empirical lens of social science and the sociology of law.

V.O. Ballano, *Law, Normative Pluralism, and Post-Disaster Recovery*,
DOI 10.1007/978-981-10-5074-9_1

However, complex public post-disaster projects for victims of major natural disasters are no ordinary projects and are governed by complicated official laws and informal normative standards of various participating groups. Stakeholders, organizations, and agencies involved in a complex PDR project—as well as their respective formal and informal networks of rules, regulations, and cultural normative standards in implementing it—are often numerous, intertwining, and competing with the state's disaster legislation to provide disaster victims with adequate housing and relocation.

This book aims to show how legal and normative pluralism matter most in the outcome of the disaster victims' long-term PDR, especially in the Philippine context. It illustrates how the overlapping legal and nonlegal normative orders of the different public and private agencies in the government PDR project can result in multiple negative unintended effects that intensify the vulnerability of typhoon victims.

1.2 The Philippine Disaster Risk Reduction Management Act of 2010

The Center for Research and Epidemiology Disasters (CRED) ranked the Philippines as one of most disaster-prone countries in the world.[1] The Philippines is also fourth in the world among countries hit by the highest number of disasters during the past 20 years according to the United Nations Office for Disaster Risk Reduction (UNISDR).[2] In terms of tropical storms, the country is hit by an average of 20 typhoons per year per as it is geographically situated in the West Pacific Basin (NDRRCC Report, 2011).

In the Philippines, 70% of the country's disasters are due to hydro-meteorological phenomena, such as typhoon and flooding, and the poor are often the most vulnerable disaster victims. Social vulnerability in a population is not evenly distributed. Some regions are more susceptible to the impacts of hazards than others based on the characteristics of the people residing within them (Cutter & Emrich, 2006, p. 102). Typhoon Ketsana, popularly known in the country as Typhoon Ondoy, is one of the most devastating typhoons that hit the Philippines in 2009 which revealed differential vulnerabilities of the Philippine population. Its impacts largely depend on the physical attributes of the place and the social characteristics of people residing in it. Some areas suffered devastating losses in life

[1] http://www.bworldonline.com/content.php?section=Nation&title=Philippines-tops-2011-list-of-countries-hit-by-disasters&id=45763.

[2] "The Human Cost of Human-Related Disasters: 1995–2015" Retrieved from http://www.unisdr.org/7B39392C-70B6-4451-8831-EAA0AA3B3130/FinalDownload/DownloadId-45F0A75073EB82DEA2E80C3CB638DFEC/7B39392C-70B6-4451-8831-EAA0AA3B3130/2015/docs/climatechange/COP21_WeatherDisastersReport_2015_FINAL.pdf.

and property compared with other places. Countries such as the Philippines—with limited economic resources, low levels of technology, poor information and skills, poor infrastructure, unstable or weak institutions, and inequitable empowerment and access to resources—have little capacity to adapt and are highly vulnerable (Penalba, Elazegui, Pulhin, & Cruz, 2012, p. 310). The urban poor, who are totally dependent on social services and usually less able to respond effectively to disasters, suffer the most during disasters (Cutter, Boruff, & Shirley, 2003; Morrow, 1999). Because they are socially and economically marginalized in society, the poor are usually mostly ignored during disaster recovery (Morrow, 1999; Tobin & Ollenberger, 1993). The people who were severely affected by Ketsana in the Philippines were urban poor residing in makeshift shanties in low-lying areas or near creeks and rivers.

Typhoon Ketsana did not only expose the differential vulnerabilities of the Philippines. It also revealed the inadequacy of the country's disaster-management law in dealing with PDR after large-scale disasters. With the unexpected loss of lives, destruction of property, and a number of people homeless, the Philippine government updated its old primary disaster-management law and enacted the Philippine Disaster Risk Reduction Management Act of 2010 (PDRRMA). Unlike the old law (PD 1566), the PDRRMA is said to be pro-active and holistic in its approach and has, for the first time, addressed the issue of PDR. The Philippine government, as well as Ketsana post-disaster responders, became optimistic with the enactment of PDRRMA that thousands of homeless disaster victims would finally receive adequate housing and resettlement. Years have passed since the passage of this law, and beneficiaries have received housing assistance from the government, non-government organizations, and foreign and local donors. Yet, in-depth qualitative studies on how the law's post-disaster provisions were implemented on the ground, given the multiplicity of actors and the plurality of legal and non-legal normative orders that surround PDRRMA, seemed missing.

1.3 The Significance and Objectives of the Book

Of the four major phases of disaster management—preparedness, response, recovery, and mitigation—recovery is often viewed by scholars as the most poorly understood and the least well-researched stage (Barton, 1969; Rubin, Saperstein, & Barbee, 1985; Schwab, 1998). Disaster-recovery studies are few and systematic comparative studies are fewer (Olshansky, Johnson, & Topping, 2006). In particular, research on attributing the success and failures of PDR projects, especially from a social science perspective, is one of the most underdeveloped research areas in disaster management. Failures of PDR and reconstruction projects are often ascribed to problems in integration (Ye & Okada, 2002), finances (Freeman, 2007), inappropriate assessment (Kennedy et al., 2008), communication and coordination (Chang, Wilkinson, Potangaroa, & Seville, 2010), inadequacies of resource procurement (Chang et al., 2010), ineffective design (Ika et al. 2012), transportation (Matsumaru, Nagami, & Takeya, 2012), corruption (World Bank, 2013), and delay (Iwai & Tabuchi, 2013; Moloney, 2014; Boen, 2006; Steinberg, 2007; Nazara & Resosudarmo, 2007; Matsumaru, Nagami, & Takeya, 2012). Thus, effective project management is often seen to play an important role to ensure that PDR and reconstruction projects are carried out successfully (Hidayat & Egbu, 2010).

Investigating the success and failure of PDR projects from the perspectives of the sociology of law and normative pluralism is apparently less explored in disaster-management research. This lack of interest in examining the legal contexts of PDR programs reflects the worldwide trend in disaster research which sidelines the social-science approach in studying the broader social forces behind disaster response and management. Thus, Deflem (2012) lamented that the literature on disaster preparedness and response are mostly written from a policy-oriented vantage point and rarely carried out in an empirical and social science fashion. Specifically, the fields of social-legal studies, including criminal justice and the sociology of law, have been relatively absent or, at least, have not been situated in the specialty area of the sociology of disasters (Deflem, 2012, p. ix). Consequently, the employment of normative and legal pluralist perspectives to evaluate PDR programs has been overlooked in disaster-management research. In particular, attributing the inadequacy or failure of the housing and relocation of a long-term PDR to a group of disaster victims to the polycentric implementation of the state's official disaster law is least explored in PDR. Also, the normative pluralist environment behind PDR projects i.e., the multiple and conflicting laws, informal rules, and cultural normative structures that implement the official national disaster law is apparently absent in post-disaster research as an important explanatory variable to account the failure of PDR projects. Yet, the quantity and quality of normative orders are crucial in the success and failure of PDR projects, especially in a multilevel government PDR project after a major calamity. The plurality of legal orders or legal pluralism surrounding a PDR project, for instance, can produce hybrid or mixed legal environments that could possibly link state, local, and non-state actors and blur the lines between and thus make coordination and implementation of the disaster programs lacking in focus, which can result in negative unintended effects (Roseveare, 2013). This pluralism would even be more complex if the numerous laws and regulations are accompanied by a multiple nonlegal cultural normative order of the various stakeholders, groups, or individuals involved in the PDR project. Thus, more unintended consequences are expected to occur which can deviate the original PDR goals of the state's PDRRMA.

This book aims to illustrate how the normative and legal pluralist structure of the long-term PDR of Typhoon Ketsana victims in a Philippines relocation site has created numerous intermediary networks governed by some dominant informal rules that compete with the official law and regulation, thus resulting in a polycentric interpretation and implementation of the PDR project goal. The brief, broad, and inadequate legal provisions of the PDRRMA regarding the long-term PDR program of disaster victims has given way to legal pluralism, i.e., the adoption of a multiple pre-existing non-disaster laws, rules, and alliances that directly pertain to the relocation of the urban poor living in danger zones under the Urban Development and Housing Act of 1992 (UDHA) and not that of disaster victims. This brief legislation also calls for the adoption of the UDHA's multiple affiliate housing laws and cultural norms of the various public and private organizations that support them. This multiplicity of cultural and informal normative orders has deviated much the original PDR goals of the PDRRMA not just in the relocation system but also in other forms of post-disaster housing assistance, whether in the form of emergency shelter cash assistance or civil society-private donor housing projects. In countries with so-called collectivist cultures that provide more value to the welfare of the in-group than the common good, to informal norms than regulations such as the Philippines. Thus, informal normative systems tend to sideline the official legal system, resulting in negative unintended consequences to a PDR project.

Applying the normative pluralist approach to examine the unintended effects of the plurality of normative systems in a social order, this book primarily investigates the "failure" of the long-term PDR for Typhoon Ketsana victims in the Southville Rodriguez Housing Project (SRHP), particularly Phase 8A in Rodriguez, Rizal. Specifically, it aims to determine the level of adequacy of the housing and resettlement program provided by the government, six years after the devastating typhoon in 2009. It analyzes how the numerous formal and informal rules that implement the housing project affect the actual PDR program of the Typhoon Ketsana victims: Whether or not the relocation in SRHP satisfactorily complied with the legal provisions of the PDRRMA and the international standards on PDR of disaster victims. Moreover, it also analyzes how normative pluralism has affected the post-disaster housing program in the Philippines under the legal framework of the PDRRMA. It assumed that the diversity of normative fields—both formal and informal—that interpreted the PDRRMA has deviated the official law's intended goals for the disaster victims' PDR program, thus providing more discretion to project officers and allowing some powerful public and private interest groups, through their social networks and informal rules, to determine the final project outcome.

The Secretary-General of the United Nations issued a document entitled Guiding Principles on Internal Displacement to guide national governments in handling internally displaced persons inside their borders last February 11 1998. This was released more than a decade ago, but little research has been done to realize the extent to which governments that relocate populations are exercising this responsibility within the context of these guiding principles (UNHCR, 2014). This book, therefore, also aspires to fill this lack of research on how governments relocate displaced people such as homeless typhoon victims using the perspectives of Sociology of disasters. In particular, it illustrates how the Philippine government implements these principles to displaced disaster victims of Tropical Storm Ketsana in the country, specifically to those in SRHP, Phase 8A, in Rodriguez, Rizal.

1.4 Evaluating Typhoon Ketsana Victims' PDR

The government's PDRRMA paints a holistic approach to disaster management that includes post-disaster rehabilitation for calamity victims. But one may ask: How is this official law being interpreted and implemented in the actual long-term PDR project involving various public and private agencies and informal groups?

After Typhoon Ketsana struck in September 2009, thousands of Filipinos, mainly the urban poor who reside in danger zones, became homeless. Local governments immediately provided typhoon victims with relief goods and temporary shelters PDR in public schools, gyms, and barangay halls for their short term PDR. Victims were housed in makeshift and temporary shelters near barangay halls while waiting for the approval of their applications for relocation and housing through the National Housing Authority (NHA), the national government's low-cost housing agency for long-term PDR. Some victims of Metro Manila and nearby Province of Rizal were eventually relocated to the government's SRHP in the remote barangay of San Isidro, Rodriguez, Rizal, after six months to a year of waiting for the approval of their housing application. With the newly enacted PDRRMA which promised a more holistic approach to disaster management, the national government assured the typhoon victims through the NHA and local government, that they would be given adequate resettlement under the BBB principle.

Seven years have elapsed since the implementation of the government's long-term PDR program to Typhoon Ketsana victims in SRHP under the PDRRMA and its affiliate laws. But beneficiaries still complained the site risks and subhuman conditions inside the resettlement sites, such as those in the SRHP. Reports in the mass media and nongovernment organization (NGO) research on site risks continued to persist. Investigations and petitions by some NGOs and religious groups indicated a greater vulnerability for disaster victims inside government-owned housing and relocation areas.[3] COHRE's research, for instance, "revealed numerous problems associated with the relocation process such as the lack of consultation with affected families; carrying out evictions and relocation before the sites are habitable, insufficient Government loans to affected families for the construction of homes; lack of livelihood opportunities for those who have been relocated because the sites are far away from Metro Manila."[4]The government's SRHP, which is supposed to be built proximate to job sites and livelihood opportunities and located in safe areas in accordance with the official law and BBB principles is actually

---

[3]A fact-finding mission by Demolitionwatch in the SRHP Phase 8B, for instance, revealed that relocatees have no security of tenure to their housing units. The relocation area lacks materials, services, and infrastructure, and the housing units are defective and made of substandard construction materials, etc. See urgent appeal to the United Nations at https://demolitionwatch.wordpress.com/2012/02/Urgent-appeal-f-2/. The religious group, Claretians, also made a similar appeal to the government to improve the lives of the relocatees in the SRHP.

[4]UN-HABITAT (2007). Forced Evictions—Towards Solutions?: Second Report of the Advisory Group on Forced Evictions to the Director of UN-HABITAT. Nairobi, Kenya: Advisory Group on Forced Eviction (AGFE) pp. 43–44.

remote from employment sites. The housing project is located in a very disaster-prone site, even more dangerous than the victims' former residential areas in "danger zones." It is also too close to the two of the country's major fault lines—the East and the West Valley Faults. Thus, the SRHP is prone to strong earthquakes. In fact, the Philippine Institute of Volcanology and Seismology (PHIVOCS) is expecting a very strong earthquake, called the "Big One" with a magnitude of 7.2, to hit Rodriguez Municipality where the SRHP is located and Metro Manila at any time.[5] Moreover, it is also reported by residents and NGOs that the housing units of SRHP were structurally weak and made of sub-standard materials.[6] Lastly, some investigations also revealed a lack of adequate social and health services and facilities in the relocation site. The entire SRHP, for instance, is said to have no adequate health and educational facilities. There are no hospitals proximate to the relocation site. Thus, one wonders: To what extent are these complaints and reports true? If true, what went wrong with the government's housing and relocation project to Typhoon Ketsana victims? What primarily caused these negative unintended effects or failures to the PDRRMA's long-term PDR in SRHP? What is the role of the multiple, polycentric, and competing laws, rules, regulations, and informal normative cultural systems of the project's various agencies and groups in determining the success or failure of the PDR goals of PDRRMA to Typhoon Ketsana victims?

1.5 Theoretical Orientation of the Book

This book deviates from the popular approach in evaluating the level of success and failure of PDR projects and applies a sociological and normative pluralist theoretical framework to account for the unintended effects of the plurality of legal and nonlegal normative systems that interpret and implement the government resettlement project to a group of Typhoon Ketsana victims in SRHP in Rodriguez, Rizal. Failures in long-term goals of PDR programs are usually attributed to poor project management. But they are rarely blamed the unintended consequences of the polycentric and multiple normative standards, both formal and informal, or what scholars call "normative pluralism,"especially the social and cultural norms that lie in between the official laws that govern the project and empirical reality.

---

[5]The West and East Valley Fault line system spans from Montalban (Rodriguez), Rizal, i.e., near the SRHP, all the way to Carmona, Cavite, and is considered the most feared event, known as the "Big One," in case it moves. It could affect millions of people and paralyze cities and towns nearby, including Metro Manila, because it is predicted to occur at any time registering a >7.0 magnitude earthquake. See interactive map at http://www.s1expeditions.com/2014/07/158-west-east-valley-fault-line.html.

[6]See "Relocation sites: the land of empty promises in the middle of nowhere." Retrieved from http://bulatlat.com/main/2015/01/14/relocation-sites-the-land-of-empty-promises-in-the-middle-of-nowhere/.

A set of formal and informal rules always guide managers in implementing projects. This book assumes that the quantity and quality of normative standards that informally support the legal goals of PDR project under the PDRRMA are crucial in minimizing the negative unintended consequences which lead to the failure of the housing project and greater suffering of disaster victims.

This section structures the theoretical orientation of the study and guides the evaluation and sociological analysis of the book on the level of success or adequacy of the long-term PDR project under PDRRMA for the housing and relocation of Typhoon Ketsana victims in the government-owned SRHP in the following areas, namely: the accessibility, affordability, and habitability of the housing units as well as the availability of educational and social services, jobs, and livelihood to disaster victims.

1.6 Areas of Success and Failure in PDR Projects

A project can be defined as a "large or important item of work, involving considerable expense, personnel, and equipment, a one-time endeavor with a specific result or end-state envisioned (Kerzner, 2003; Benator & Thumann, 2003), with a proper plan (Dhillon, 2002)" (as cited in Ismael et al., 2005). A PDR project is often undertaken by a variety of stakeholders, agencies, and individuals. Thus, to ensure that decisions and participation of the various agencies and stakeholders are in accordance with goals of the PDR program, the project-management approach is usually employed in managing many post-disaster humanitarian relief and recovery projects (LaBrosse, 2007). The management approach usually determines a project's success in terms of scope, time, and cost. Project-management literature suggests that project success revolves around planning, defined goals and objectives, top management support, and financial support (Marchewka, 2006; Hughes & Cotterell, 2002). Most PDR projects that apply the management approach also recognized these factors as a key "areas of success."

Planning is said to be crucial to the success of any PDR project. The most common attribution of PDR failures is a lack of proper planning. Disaster planning is said to be important as an effective human intervention in achieving sustainable community development when confronted with natural disasters (Ge, Gu, & Deng, 2010). In particular, recovery and reconstruction planning as an important component of disaster-management systems represents a comprehensive response and proactive adjustment in disaster recovery (Turner et al., 2003). Specifically, the power of pre-event planning is also considered as crucial to the success or failure of a PDR project. The literature shows that such planning improves outcomes at the local level (Oliver-Smith, 1991; Berke, Kartez, & Wenger, 1993). To some scholars, careful planning, as well as implementation, are highly important for the success of post-disaster reconstruction. They argued that PDRs should be well defined, planned, and implemented in stages (Roosli, Vebry, Mydin, & Ismail,

2012) and that their existing tools or new tools must be adapted to attain their goals; otherwise, further vulnerabilities in a disaster-affected community can occur (Chang, et al., 2010).

Another area that is crucial to the success of PDR projects is the participation of the beneficiaries in the planning and implementing the programs and effective management of them (Baroudi & Rapp, 2010). In fact, community involvement is usually a compulsory component for funding organizations (Davidson et al., 2007, p. 2) to stress that beneficiaries' participation is crucial to the success of any PDR program. Thus, Oliver-Smith (1991), presenting case-study materials on the problem of resettling a population after disasters in Turkey, Iran, and Peru, argued that public participation, aside from the appropriateness of the site, layout, and housing of the relocation, is crucial for the success of post-disaster resettlement projects. The public input in terms of communal involvement is said to be an important component of a successful PDR project. Moreover, an effective project management can also be crucial for the success of PDRs. Poor management can lead to disastrous effects to PDR projects. It also plays a significant role to ensure the successful completion of reconstruction projects (Baroudi & Rapp, 2010; Hidayat & Egbu, 2010). Munns and Bjeirmi (1996), for instance, stressed that effective project-management techniques would contribute to the achievement of the goals of projects (Ismael et al., 2005).

Other scholars point to the adequacy of the physical environment for disaster victims, i.e., the proximity of the relocation to urban jobs and livelihood, and the empowerment and independence of beneficiaries from settlement agencies as crucial in success factors. Coburn et al. (1984), for instance, identified the physical environment of the new settlement, relationship to the old village, and capacity of the community to develop itself as factors critical in determining the success or failure of a resettlement project. Most beneficiaries, especially those from urban areas, returned to their old location to resume their informal jobs and livelihood if the resettlement area is remote and far from livelihood opportunities. Thus, in-city relocation would more likely result in successful resettlement than out-city relocation. In Japan, for instance, Imura and Shaw (2009) noted that in the Iwate prefecture of Japan after a tsunami, village people were relocated to the mountainside but soon returned to their previous location to resume their livelihood in fisheries. Other studies have also pointed out greater autonomy and empowerment of the relocated beneficiaries as crucial for the success of PDR projects. Michael (1988) argued that successful resettlement schemes should result in a transfer of responsibility from settlement agencies to the settlers themselves.

Finally, some authors have used a multivariate approach to identify the success factors in PDR projects. Moe and Pathranarakul (2006), for instance, in their case study of a post-disaster reconstruction project in Thailand, considered the following factors as areas for success and failures: effective institutional arrangement; coordination and collaboration; supportive laws and regulations; effective information management system; competencies of managers and team members; effective consultation with key stakeholders and target beneficiaries; effective communication mechanism; clearly defined goals and commitments by key stakeholders;

effective logistics management; and sufficient mobilization and disbursement of resources (Moe & Pathranarakul, 2006). A case study of Kim and Choi (2013) on flood rebuild projects also used a multivariate approach and identified five factors as critical for the success and failure of a government PDR: clear project execution plan; improvement in design management; enhancement of coordination at the plan–design-construction interface; elimination of the vicious cycle for improving project performance; and rapid evaluation of a contractor's qualification.

Although community participation, proper planning and implementation, and other factors are crucial for the success of PDR projects and perhaps considered as proximate causes, they are not, however, the ultimate cause and are not sufficient to account for the inadequacies and numerous negative unintended effects that occurred to a government-initiated post-disaster housing and relocation in the Philippines. All projects and their social contexts are different. Thus, the success criteria must be determined for each project rather than just using a standard set of success criteria (Cooke-Davies, 2002).

Disasters are often treated as external to the law in project management. Yet social processes in disaster mitigation, response, and recovery include mobilization of the law by people and organizations (Sterett, 2013). Disasters as social phenomena are embedded in both legal and non-legal normative orders. Government disaster projects are particularly juridified and are usually embedded in multiple organizational contexts (Sydow et al., 2004). Thus, the conceptualizations of the success criteria for government post-disaster projects must take into account the multi-level and legal and bureaucratic nature of a resettlement project as well as the plurality of informal norms that surround the official laws during the implementation phase. National and local governments—with their multiple networks of stakeholders such as public and private agencies, non-profit organizations, community groups, and individuals—can create numerous unintended consequences. This plurality of participating groups in the project can lead to stakeholder issues (Baroudi & Rapp, 2014; Haigh & Sutton, 2012; Walker et al., 2014), resource challenges (Chang et al., 2010; Chang, Wilkinson, Seville, & Potangaroa, 2012), capability issues (Crawford et al., 2012), long-term reliability concerns (Hayes & Hammons, 2000), coordination issues (Drabek & McEntire, 2002; Quarantelli, 2000), and even corruption issues (World Bank, 2013).

In a highly complex, graft-ridden, and bureaucratic real estate–regulatory system, such as that of the Philippines, the quantity and multiplicity of laws, rules, and informal normative orders for the enforcement a PDR project are crucial in the success criteria of government projects. Legal pluralism and lack of a coordination mechanism for actors and organizations in the shelter sector can create regulatory barriers to providing emergency and transitional shelter after disasters (IFRC, 2015). It is an undeniable fact that public projects, such as PDRs, must follow strictly the multiple national and local legal provisions and their implementing rules and regulations (IRRs); otherwise, they can be nullified by the courts and the legal officers, and their private collaborators involved in illegalities can be prosecuted for violations of the articles of the Anti-Graft and Corrupt Practices Act (R.A. 3019). However laws, policies, and institutional arrangements for disaster management are significantly shaped by the Philippines' highly decentralized governance system in the country as prescribed by the 1987 Constitution and the Local Government Code of 1991 (Sherwood et al., 2015). In the absence of a "one-stop" disaster-management agency and a coherent set of laws or Magna Carta on disaster management in the Philippines, negative unintended consequences to a PDR project can occur. Thus, the simplicity and cohesiveness of laws and informal norms in the recovery projects, as well as their legal and nonlegal normative systems, can be critical in determining the success or failure of government PDR projects.

Normative pluralism, both in the legal and cultural normative orders, can produce hybrid or mixed normative environments which can result in multiple unintended deviations from the original objectives of the PDR project. Moreover, pluralistic and bureaucratic normative structures in a government disaster-recovery program can also lead to delays and greater costs to PDR projects and encourage managers to behave unethically in project implementation, thus resulting in more negative impacts to the project's outcome. In an environment of legal pluralism, the "Rule of Law," with its checks or balances, is said to erode, and the accountability of public officers becomes problematic (Grenfell, 2006). When government rules and regulations for business are complex and highly bureaucratic, individuals and groups can resort to informal channels and offer "grease money" and similar dishonest dealings to expedite transactions (Ballesteros, 2000). Once corrupt patterns are established, they tend to perpetuate to avoid uncertainty (Seleim & Bontis, 2009). No less than the World Bank acknowledged the negative consequences of corruption to PDR projects (World Bank, 2013).

The Philippines is known for corruption, red tape, delays, and complex bureaucratic requirements in doing business in the country (Ballano, 2016), particularly about housing and real-estate transactions such as the establishment of a housing and relocation projects (Valte, 2000). Heavy, costly, and complex bureaucratic burdens in business often breed corruption. Thus, the Philippines is consistently among the most corrupt countries in the world with high corruption index in government transactions and projects according to Transparency International (TI) annual surveys. The World Bank underscores corruption as a major obstacle in the implementation of PDR projects. Thus, the Philippines, with its high level of corruption in real-estate projects due to bureaucratic burdens and plurality of PDR agencies, organizations, and groups—as well as pluralism in the normative systems that implement PDR projects—illustrates a unique case in understanding the success factors in public post-disaster housing projects.

This book also explores the influences of some negative or corrupt cultural normative systems in the implementation of low-cost housing laws and the PDRRMA's PDR legal objectives for Typhoon Ketsana victims in SRHF (Fig. 1.1). This book evaluates how the plurality of legal and informal normative systems affect the post-disaster housing and relocation of Typhoon Ketsana victims according to the criteria of the United Nations' Committee on Economic, Social,

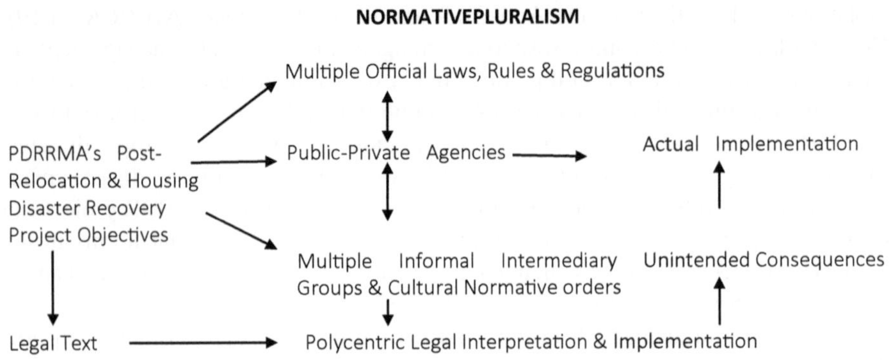

Fig. 1.1 Conceptual framework of the book

Fig. 1.2 Overview of the entire SRHP Relocation Area showing some housing units of Phase 8, 8A and Phases 8B and 8C as shown in the remote background. Note that the relocation is situated in a valley surrounded by mountains where quarrying and mining activities occur. *Source* Author

and Cultural Rights (ICESCR) for an adequate PDR of disaster victims under the "build back better" (BBB) principle as embedded in PDRRMA of 2010, more specifically on the following key areas: (1) Security of tenure, (2) suitability of location, (3) accessibility, (4) habitability, (5) affordability, and (6) availability (Carver, 2011). It also uses the criteria of Principle 18 of the United Nations' Guiding Principles on Internal Displacement which provides that all internally-displaced persons must have the right to an adequate standard of living and safe access to (a) essential food and potable water; (b) basic shelter and housing; (c) appropriate clothing; and (d) essential medical services and sanitation. Special efforts should be made to ensure the full participation of women in the planning and distribution of these basic supplies.[7]

[7]http://www.brookings.edu/ ~ /media/Projects/idp/GPEnglish.pdf.

Altogether, this book examines how these criteria for an adequate post-disaster program are realized in the SRHP, Phase 8A relocation site, through the PDR provisions of the PDRRMA and its manifold affiliate laws and pre-existing informal normative relations among the many government and private organizations involved in the post-disaster housing project for Typhoon Ketsana victims in Rodriguez, Rizal.

1.7 The Case Study

The research on which this book is based is a case study of Typhoon Ketsana victims relocated at the government-owned SRHP in the Municipality of Rodriguez, Rizal. The fieldwork was performed by the author intermittently from June to October 30, 2012, and additional interviews and documentary data were added in 2013 and 2014. Data-collection method included direct personal observation and unstructured interviews with 31 Typhoon Ketsana victims in SRHP who were relocated in 2014. The key informants were chosen using the snowball sampling. They include six municipal officials and housing officers, two buyers of relocation houses, three barangay officials, two NHA provincial officers, one structural engineer, one licensed broker, one Land Registration Authority (LRA) official, three mason–carpenters, and thirty-one residents or Typhoon Ketsana victims. Interviewing the Ketsana informants was conducted during weekdays in the relocation area. Thus, almost all of the key informants–residents (25 out of 31) were women because most of the husbands were away working. The Philippines is still a highly patriarchal society with a traditional gender relation despite a significant development in gender equality. The husbands are still considered as breadwinners and the wives as housekeepers. Most surveys estimate that 70% of Filipino women are still housekeepers, especially those in the working class. In SRHP, 90% of the wives are housekeepers. That is why most of the key informants were women.

Aside from structured and unstructured interviews, focused group discussions (FGDs) were also conducted to understand the current problems of the beneficiaries in the relocation area and to clarify issues raised by informants during the personal interviews. Twenty beneficiaries or Ketsana victims resided in Phase A of SRHP and eleven beneficiaries lived in the neighboring Phase 8B. Moreover, newspaper reports, research articles, government documents, and audit reports on disaster management and government-housing programs as well as the personal observations of the author were also used in the research to triangulate the data.

1.8 The Research Site

1.8.1 A Brief History of SRHP

On January 6, 1999, the former Philippine President Joseph "Erap" Estrada signed the Executive Order (E.O.) No. 65, amending E.O. No. 54, which created the Pasig River Rehabilitation Program (PRRC) whose main objective is to clean and rehabilitate the Pasig River and to relocate the urban poor living or near the river "in an environment free from pollution and unsanitary conditions."[8] To finance this program, the government secured a $100- million loan from Asian Development Bank (ADB).[9] The PHP 2.8 billion peso mass-housing project called the Kasiglahan Housing Project or the New San Jose Plains, popularly known as "Erap City" in the mountains of Rodriguez (Montalban), Rizal, became one of the main destinations of the urban poor affected by the government's rehabilitation of the polluted Pasig River.

According to HPDO, the SRHP is originally part of the master plan of the Erap City project. This project was first conceived during the administration of the former Philippine President Fidel V. Ramos. Its master plan was originally prepared by Palafox Associates. The first phase of the project was originally conceived to cover only 600 hectares. The 400 hectares will be alotted for the Suburban area and the other 200 hectares for San Jose Plains which includes the Kasiglahan Village I (KV1) project. When Joseph Estrada became President, this Kasiglahan housing project was hugely expanded to accommodate as many as 1.2 million residents from Metro Manila, including the urban poor who were relocated because of slum upgrading and informal settlers who were displaced due to the rehabilitation of Pasig River. The expanded Kasiglahan project became known as the Erap City housing project became the centerpiece of the Estrada administration's housing program and was touted as the "first-ever well-planned socialized housing city undertaken in the country."[10] Its land area covers 5 barangays of Rodriguez, Rizal —San Isidro, Mascap, San Jose, Puray, and Macabud—and stretches across 2500 hectares of land in the municipality.[11]

Erap City, however, sits on disaster-prone areas of Rodriguez, Rizal. The Philippine Investigative Journalism (PCIJ) claimed that the development of Erap City started without an Environmental Compliance Certificate (ECC). Thus, a notice of violation was issued by the government's Environmental Management Bureau on February 24, 2000, 2 days after development of the suburban area began. The firm paid a PHP 50,000 fine on June 2, 2000; applied for an ECC; and got one on June 8, 2000. Despite the issuance of the ECC to Erap City, the fact

[8]http://www.chanrobles.com/executiveorderno54estradaprrcjanuary061999.html#.VQPS-47A3Mw.

[9]Because a national budget has been set aside for the construction of government relocation and housing projects, a portion of the ADB loan is intended for the social services and livelihood programs of relocatees affected by Pasig River rehabilitation. However, disgruntled relocatees and members of the KV1 Action Group asked President Gloria Macapagal–Arroyo's government to probe the anomalous PHP 2.116 billion budget of the Pasig River Rehabilitation Commission (PRRC) for its resettlement projects. They alleged that the Estrada government wasted the huge amount of loan from the Asian Development Bank (ADB), which is part of the overall PHP 7.9 billion budget for the river's rehabilitation (http://bulatlat.com/archive1/009erap_city.htm).

[10]http://pcij.org/stories/2000/erapcity.html.

[11]See Fig. 1.2.

remains that its relocation site is situated in disaster-prone barangays. Some parts of this grand relocation site, for instance, are situated in the flood plains of Barangay San Jose and are close to the three watershed areas—Angat Dam, which supplies 97% of Metro Manila's water; Marikina Dam; and the La Mesa Dam of Novaliches —and thus are prone to flooding. Maps from The Philippine Institute of Volcanology and Seismology (PHIVOLCS) also showed that barangays Macabud and San Jose are situated in the upper block of the West Valley Fault lines. Thus, some parts of Erap City are prone to surface rupture in case of strong earthquakes, thus further endangering the lives of relocatees.[12] In fact, 60 houses of Kasiglahan I (KV1) sit directly at the top of the West Valley fault line and thus have been recommended by PHIVOLCS for demolition.[13]

The SRHP, being part of the Erap City complex, shares these geographical hazards. However, NHA-accredited developers of Erap City were able to evade the government regulation against any construction of a housing project in areas with environmental hazards. After completing Kasiglahan Village 1 (KV1), they continued to expand the Erap City housing project during the administration of President Gloria Arroyo and constructed the SRHP using the previous Environmental Clearance Certificate (ECC) they obtained for the KV1 relocation area. Thus, the construction of houses of SRHP's Phase 8 and 8A commenced before Typhoon Ketsana struck the country in October 2009. These two sites were originally intended for disaster victims of Taytay and Rodriguez (Montalban), Rizal, before Typhoon Ketsana. However, when Ketsana hit the country, the relocation site was, instead, used to accommodate more Ketsana victims of Rodriguez and other neighboring areas. From 2010 onward, SRHP expanded its area and opened new phases to accommodate more disaster victims and displaced informal settlers from various places of Metro Manila and neighboring towns of Rizal such as Taytay, San Mateo, Marikina, and Montalban (Fig. 1.3).[14]

Most of the Rodriguez Typhoon Ketsana victims were relocated in the SRHP Phases 8, 8A, 8B, and 8C. These phases are situated in the adjoining barangays of San Isidro and San Jose, Rodriguez, Rizal The NHA-accredited developers of SRHP are Baque Corporation, Gateway Sand Builders Inc., and San Jose Builders Inc.

The SRHP is one of the largest relocation sites in the country as well as in Erap City. As shown in Table 1.1, the entire SRHP had 20,477 housing units as of October 2012. Construction is ongoing. The NHA is expecting that the low-cost houses of the relocation will reach as many as 60,000 units.

Table 1.1 showed that the SRHP site is composed of various socialized housing projects intended for victims of man-made and natural disasters in Metro Manila and the province of Rizal. Typhoon Ketsana victims were relocated in Phase 8A and 8B. Phase 8A is the smallest phase of the SRHP with only 1007 housing units. The neighboring Phase 8B has 7,541 houses, while SRHP Phase 8C which

[12]http://bulatlat.com/archive1/009erap_city.htm.

[13]http://pcij.org/stories/2000/erapcity.html.

[14]http://inclusivemanila.net/southville-8/.

Fig. 1.3 An aerial view of some portions of Erap City showing the Kasiglahan Village I (KV1) and SRHP relocation areas. *Source* http://inclusivemanila.net/southville-8-rodriguez-rizal/

Table 1.1 Profile of the SRHP in terms of location and number of sites, housing units, and developers

Name of relocation	Number of phase	Barangay location	Number of housing units	Developer
the SRHP	0	San Isidro	1750	Baque Corp.
the SRHP Phase 8A	1	San Isidro	1007	Gateway Sandbuilders Inc.
the SRHP Phase 8B	1	San Isidro	7541	Baque Corp.
the SRHP Phase 8C	1 (6 sites)	San Jose	10,179	San Jose Builders Inc.
Total			**20,477**[a]	

[a]As of November 2012. *Source* NHA Kasiglahan Village 1 Office

is located in Barangay San Jose, has 10,179 units. San Isidro barangay officials estimated the residents of the entire SRHP were approximately 40,000 as of November 2013 (Fig. 1.4).

1.8.2 *SRHP as a Resettlement*

The SRHP, as part of the conglomeration of relocation areas called the Erap City, follows the structure of housing programs of the national government administered by the NHA. The NHA classifies the government's socialized housing programs

Fig. 1.4 The research site: SRHP Phase 8A relocation site. *Source* Author

into 5 types: (1) resettlement, (2) slum upgrading, (3) sites and services, (4) completed core housing, and (5) medium-rise housing.[15] The SRHP is a resettlement housing program. It "involves the acquisition and development of large tracts of raw land to generate serviced lots and/or housing units for families displaced from sites earmarked for government infrastructure projects and those occupying danger areas such as waterways, esteros, and railroad tracks" (Ballesteros & Egana, 2012, p. 13). Although resettlement programs of the NHA are primarily intended for the urban poor and informal settlers under the UDHA, they also accomodate disaster victims under the implied legal provisions of the PDRRMA. By method, it is a developer-constructed type (completed core housing), and by location it is an in-city relocation site for residents of Rodriguez, Rizal, but an off-city site for beneficiaries from Metro Manila and nearby cities and towns.

Conversely, a government resettlement is classified by law by method and by location. By method, a relocation can be either (1) a Completed Housing-Resettlement Project or a Developer-Constructed Project; (2) a Home Material Loan Project or an Incremental Housing Project; or (3) an LGU-NHA joint venture scheme or a RAP-LGU. Completed housing resettlement projects that were started in 2004 and have become the dominant mode of relocation and housing in the Philippines allow the developer to offer developed resettlement sites with completed housing units to the community associations, which in turn select the site and execute a formal agreement with the developer on the proper endorsement of the Local Inter-Agency Committee (LIAC) and the NHA of the selected site. The NHA provides the loans to beneficiaries for the acquisition of the housing units, and the loan proceeds are paid to the developer on delivery of the housing unit and formal acceptance by the individual beneficiary. In incremental housing, each beneficiary or family receives a loan (PHP 75,000 in 2008)

[15]For a more detailed explanation of these types of resettlement programs, see Ballesteros and Egana (2012) at http://www.dbm.gov.ph/62C97C27-20B5-4CB0-9BD2-8286EDF12907/Final Download/DownloadId-802ACD5E5502BED15120A5F66A855713/62C97C27-20B5-4CB0-9B D2-8286EDF12907/wp-content/uploads/DBM%20Publications/FPB/ZBB-2012/e.pdf.

from the NHA, payable in 30 years, for the construction of their homes in a NHA or government-owned relocation site, whereas in the LGU-NHA joint venture, which is usually performed outside of Metro Manila, the relocation and housing project becomes a shared undertaking between the LGU and the NHA. The LGU usually contributes the land, identifies the beneficiaries, and takes care of the operation and management of the relocation site, while the NHA contributes funds to the development of the site and construction of housing units and provides technical expertise for the preparation of project plans and the formulation of policies and guidelines for implementation of resettlement projects (Ballesteros & Egana, 2012, pp. 6–7).

A resettlement project by location can be in-city or an off-city site. An in-city site is one where the relocation and housing are located within the same Local Government Unit (LGU), whereas an off-city is one whose location is considered distant (possibly 20–30 km from the original settlement) and outside the administrative boundaries of the LGU (Ballesteros & Egana, 2012, p. 18). By location, the SRHP is an in-city relocation for the disaster of the municipality of Rodriguez, Rizal. It is situated in Barangay San Isidro and San Jose, two of the biggest barangays of Rodriguez, Rizal. To the disaster victims of Metro Manila and neighboring towns and cities, SRHP is considered an off-city relocation because it is located outside their jurisdictions.

The SRHP Phase 8A is in Sitio Tanag, Barangay San Isidro, Rodriguez, Rizal. Together with SRHP Phase 8, it was constructed before Typhoon Ketsana hit the country in September 2009. It was one of the earliest phases of the SRHP that was constructed under the adminsitration of former President Gloria Arroyo. When Ketsana hit the country, the SRHP was expanded to accomodate thousands of homeless typhoon victims. Since then, the SRHP has been expanded to shelter other disaster victims, as well as urban poor whose homes have been demolished by local governments. SRHP then became one of the largest resettlement areas in Erap City. Therefore, the SRHP is not only intended for Typhoon Ketsana victims but also for victims of other natural and man-made calamities, specifically for those who experienced demolition or forced relocation by the local government. Informal settlers who reside the danger zones such as rivers, creeks, or near railways are usually relocated by LGUs to safer areas. Every phase of SRHP has different lot sizes and house floor areas. The first phase of SRHP has a total land area of 8.5 hectares, and the average housing unit has a 40 m^2/lot area and with a floor area of 22 m^2.

1.9 The Roadmap of the Book

The overall objective of this book is to evaluate the extent of conformity of the SRHP's housing and relocation for Typhoon Ketsana's victims in Rodriguez, Rizal, with the PDR legal goals of the PDRRMA, and the UN's PDR treatises in the following four key areas of the project—namely, the suitability of the relocation site; security of ownership, affordability, and habitability of the housing units; availability of jobs and livelihood; and the adequacy of the social facilities and services. Typhoon

Ketsana victims were relocated in the SRHP's Phase 8A. To achieve this, the chapters are arranged to provide readers an overview of the book's content.

Chapter 1 (Introduction) provides an overview of the case study on which it is based. It discusses the background, research objectives, significance of the study, overall theoretical framework, method, and gives a short history and background of the research site. It provides a brief description of the low-cost housing and relocation system in the Philippines. It aims to describe the plurality of laws, informal rules, cultural norms, and legal and quasi-legal orders in Philippine relocation and low-cost housing system and how they can affect the interpretation and implementation of the PDR program of the PDRRMA for Typhoon Ketsana victims.

Chapter 2 provides the general theoretical framework of the book. It reviews broadly the history and trends in the sociology of disasters and explains why the sociological normative pluralist perspective is appropriate for analyzing the unintended effects of enforcing PDR projects in disaster-prone countries with multiple legal and social normative systems, rigid bureaucratic regulations, and high red-tape and corruption indices such as the Philippines.

Chapter 3 examines the holistic approach and PDR standards of the Philippine Disaster Risk Reduction Act (PDRRMA) of 2010. The PDRRMA's lack of legal provisions that guide the actual post-disaster housing for Typhoon Ketsana victims resulted in the reliance of the PDR managers of the SRHP on the existing pluralism in laws and informal normative systems that govern real-estate business. With the Philippines' complex real-estate regulation system and personalistic cultural values of the stakeholders, serious deviations of the intended recovery goals of PDRRMA would likely occur, increasing the suffering of typhoon victims.

Chapter 4 discusses the legal pluralism in the Philippine housing and relocation program as well as the housing itself. It provides an overview of the complexity of the formal and informal laws, rules, and regulations and the diverse groups that enforce them. It also specifies the laws governing the four important criteria on adequate relocation and housing as envisioned by the PDRRMA and international conventions: suitability, accessibility, security of tenure, affordability, habitability and availability of basic services, facilities, and livelihood.

Chapter 5 closely examines the SRHP in Barangay San Isidro, Rodriguez, Rizal, and evaluates in detail its suitability and accessibility to disaster victims in the light of the legal provisions of the PDRRMA and other related laws and international conventions on adequate resettlement for disaster victims.

Chapter 6 investigates the security of tenure of buyers or beneficiaries to their housing units in SRHP's housing project as well as the affordability and habitability of the houses built inside the resettlement. It highlights the overpricing of the land and housing units as well as the weak construction of the house and poor quality of building materials used by developers to maximize profit, thus sidelining the official law's provision on building disaster-resilient homes and communities for disaster victims.

Chapter 7 examines why a sustainable livelihood program is absent in SRHP Phase 8A, which is contrary to what the PDRRMA and other official laws say on holistic PDR. It also analyzes the contentious issue of having a livelihood program in a national relocation site, such as the SRHP, where the local government is wrestling for control over the relocation from a powerful national corporation such as the NHA.

Finally, Chapter 8 investigates the availability of basic social services and facilities in the relocation area. Specifically, it will show the lack of social services and sustainable livelihood opportunities in SRHP. Moreover, it will provide some sociological analyses on the responsibility of the NHA, the developer, and the local government of Rodriguez, Rizal, for this lack of social services and facilities. It will also examine the problems of electricity, water, health, lack of access to hospitals and educational facilities, as well as the limited religious facilities inside the housing project.

References

ADB. (2015). *Proceedings of the regional knowledge forum on post-disaster management*, October 20–21, 2015. Asian Development Bank. Accessed February 29, 2016.

Ballano, V. (2016). *Sociological perspectives on media piracy in the Philippines and Vietnam*. Singapore: Springer Science+Business Media Singapore.

Ballesteros, M. (2000). Land Use Planning in Metro Manila and the Urban Fringe: Implications on the Land and Real Estate Market. Discussion Paper Series No. 2000–20

Ballesteros, M., & Egana, J. (2012). *Efficiency and effectiveness review of the National Housing Authority (NHA) resettlement program*. http://www.dbm.gov.ph/62C97C27-20B5-4CB0-9BD2-8286EDF12907/FinalDownload/DownloadId-802ACD5E5502BED20A5F66A855713/62C97C27-20B5-4CB0-9BD2-8286EDF12907/wp-content/uploads/DBM%20Publications/FPB/ZBB-2012/e.pdf. Accessed March 4, 2015.

Baroudi, B., & Rapp, R. R. (2014). Stakeholder management in disaster restoration projects. *International Journal of Disaster Resilience in the Built Environment, 5*(2), 182–193.

Barton, Allen H. (1969). *Communities in disaster: A sociological analysis of collective stress situations*. Garden City, New York: Doubleday & Company Inc.

Benator, B., & Thumann, A. (2003). *Project management and leadership skills for engineering and construction projects*. United States of America: The Fairmont Press Inc.

Berke, P., Wenger, D., & Kartez, J. (1993). Recovery after Disaster: Achieving Sustainable Development, Mitigation and Equity. *Disasters, 17*(2), 93–109.

Boen, T. (2006). Building a safer Aceh, reconstruction of houses, one year after the Dec. 26, 2004 Tsunami. In the 40th Anniversary of Trisakti University, "Answering the Challenges in Today's Civil Engineering," January 26, 2006.

Carver, R. (2011). Is there a human right after a disaster? *Environmental Hazards, 10*(20122), 232–247. http://www.humanitarianinfo.org/iasc/downdoc.aspx?docID=4503&type=pdf. Accessed December 14, 2012.

Chang, Y., et al. (2010). Resourcing challenges for post-disaster housing reconstruction: A comparative analysis. *Building Research & Information, 38*(3), 247–264.

Chang, Y., Wilkinson, S., Seville, E., & Potangaroa, R. (2012). Changes in resource need for post-disaster reconstruction: A longitudinal study in China. *Building Research and Information, 40*(3), 327–336.

Coburn, A. W., et al. (1984). Reconstruction and resettlement 11 years later: A case study of Bingöl Province, Eastern Turkey. In S. Schupisser & J. Studer (Eds.), *Earthquake relief in less industrialized areas* (pp. 49–58). Rotterdam: A. A. Balkema.

Cooke-Davies, T,. J. (2002). Towards improved project management practice, PhD thesis, Leeds Metropolitan University.

Crawford, L.et al. (2012). *Building capability for disaster resilience*. Paper presented at the 28th Annual Conference of Association of Researchers in Construction Management (ARCOM), September 2012, Edinburgh, Scotland.

Cutter, S. L., Boruff, B. J. & Shirley, W. L. (2003). Social Vulnerability to Environmental Hazards. *Social Science Quarterly, 84*, 242–261. doi:10.1111/1540-6237.8402002

Cutter, S. L., & Emrich, C. T. (2006). Moral hazard, social catastrophy: The changing face of vulnerability along the hurricane coasts. *The Annals of the American Academy, 604*, 102–112.

Davidson, C. H., et al. (2007). Truths and myths about community participation in post-disaster housing projects. *Habitat International, 31*(1), 100–115.

Deflem, Matthieu (Ed.). (2012). *Disaster, hazards, and law. Sociology of crime, law and deviance* (Vol. 17). United Kingdom: Emerald.

Dhillon, B. S. (2002). *Engineering and technology management tools and applications* (1st ed.). Boston, London: Artech House Publishers.

Drabek, T. E., & McEntire, D. A. (2002). Emergent phenomena and multiorganizational coordination in disasters: Lessons from the research literature. *International Journal of Mass Emergencies and Disasters, 20,* 197–224.

Flatt, V. B., & Stys, J. J. (2013). Long term recovery in disaster response and the role of non-profits (April 22, 2012). *Oñati Socio-Legal Series, 3*(2). Available at SSRN: http://ssrn.com/abstract=2119879. Accessed February 1, 2015.

Freeman, P.K. (2007). Allocation of post-disaster reconstruction financing to housing. *Building Research & Information, 32*(5), 37–41 (January 2013).

Ge, Y., Gu, Y. & Deng, W. (2010). Evaluating china's national post-disaster plans: The 2008 Wenchuan Earthquake's recovery and reconstruction planning. *International Journal of Disaster Risk Science, 1*(2): 17–27. doi:10.3974/j.issn.2095-0055.2010.02.003

Grenfell, L. (2006). Legal pluralism and the rule of law in Timor-Leste. *Leiden Journal of International Law, 19*(2006), 305–337.

Haigh, R., & Sutton, R. (2012). Strategies for the effective engagement of multi-national construction enterprises in post-disaster building and infrastructure projects. *International Journal of Disaster Resilience in the Built Environment, 3*(3), 270–282.

Harris, N. R. (2010). Discretion and expediency in the enforcement of planning controls. *Town Planning Review, 81*(6), 675–700.

Hayes, P. F., & Hammons, A. (2000). *Disaster recovery project management*. Paper presented at the Industry Applications Conference, October 8–12, 2000, Rome.

Hidayat, B. & Egbu, C. O. (2010). A literature review of the role of project management in post-disaster reconstruction, in: 26th Annual ARCOM Conference, 6-8 September 2010, Leeds, UK. http://usir.salford.ac.uk/10144/

Hughes, B. & Cotterell, M., (2002), Software Project Management (2nd edition). New Jersey: Wiley & Sons.

IFRC (2015). How law and regulation supports disaster risk reduction Haiti case - study report. Geneva: Haiti Red Cross Society and the International Federation of Red Cross and Red Crescent Societies. Available at http://www.ifrc.org/Global/Publications/IDRL/country%20studies/HAITI%20DRR%20Report.pdf

Ika, L. A. (2012). Project Management for Development in Africa: Why Projects Are Failing and What Can Be Done About It. Proj Mgmt Jrnl, *43*, 27–41. doi:10.1002/pmj.21281

Imura, M., & Shaw, R. (2009). Challenges and potentials of post-disaster relocation. *Asian Journal of Environment and Disaster Management, 1*(2), 199–221. doi:10.3850/S1793924009000029

Ismael, D. et al. (2005). A review of project management for post-disaster reconstruction project: From International NGOs (INGOs) Perspective. *Research in Civil and Environmental Engineering*, 2(04), 199–215. Retrieved from https://www.academia.edu/10168784/A_REVIEW_OF_PROJECT_MANAGEMENT_FOR_POSTDISASTER_RECONSTRUCTION_PROJECT_FROM_INTERNATIONAL_NGOs_INGOs_PERSPECTIVE

Iwai, T., & Tabuchi, S. (2013). *Survey: Housing projects delayed for more than 10,000 evacuees*. The Asahi Shimbun.

Kennedy, J., et al. (2008). The meaning of "build back better": Evidence from post-tsunami Aceh and Sri Lanka. *Journal of Contingencies and Crisis Management, 16*(1), 24–36.
Kerzner, H. (2003). *Project management: A systems approach to planning, scheduling, and controlling eight edit*. New Jersey, USA: Wiley.
Kim, K., & J. Choi (2013) Breaking the vicious cycle of flood disasters: goals of project management in post-disaster rebuild projects. *International Journal of Project Management, 31*, 147–160.
LaBrosse, M. (2007). Project management's role in disaster recovery. *Leadership and Management in Engineering, 89*(July), 1.
Linda M. PeÃ±alba, Dulce D. Elazegui, Juan M. Pulhin & Rex Victor O. Cruz, (2012). Social and institutional dimensions of climate change adaptation, *International Journal of Climate Change Strategies and Management*, *4*(3), 308–322. https://doi.org/10.1108/17568691211248748
Lloyd-Jones, T. (2006). *Mind the Gap! Post-disaster reconstruction and the transition from humanitarian relief.* Summary report for RICS by the Max Lock Centre at the University of Westminster.
Marchewka, J. T. (2006). Information Technology Project Management. New Jersey: John Wiley & Sons.
Matsumaru, R., Nagami, K., & Takeya, K. (2012). Reconstruction of the Aceh Region following the 2004 Indian Ocean tsunami disaster: A transportation perspective. *IATSS Research, 36*(1) 11–19. https://doi.org/10.1016/j.iatssr.2012.07.001
Michael, A. J. (1988). Effects of three-dimensional velocity structure on the seismicity of the 1984 Morgan Hill, California, aftershock sequence. *Bulletin of the Seismological Society of America, 78,* 1199–1221.
Mileti, Dennis. (1999). *Disasters by design: A reassessment of natural hazards in the United States*. Washington, D.C.: Joseph Henry Press.
Moe, T. L., & Pathranarakul, P. (2006). An integrated approach to natural disaster management: Public project management and its critical success factors. *Disaster Prevention and Management, 15*(3), 396–413.
Moloney, A. (2014). *Haitians still homeless, "suffering in despair" 4 years after the quake-Amnesty* (pp. 4–6). Thomson Reuters Foundation.
Morrow, B. H. (1999). Identifying and Mapping Community Vulnerability. *Disasters*, 23, 1–18. doi:10.1111/1467-7717.00102
Munns, A., & Bjeirmi, B. (1996). The role of project management in achieving project success. *International Journal of Project Management*, *14*(2), 81–87.
NDRRCC Report (2011). National Disaster Risk Reduction and Management Plan (NDRRMP) 2011–2028. Available at http://www.adrc.asia/documents/dm_information/Philippines_NDRRM_Plan_2011-2028.pdf
Nazara, S., & Resosudarmo, B. P. (2007). Aceh-Nias reconstruction and rehabilitation: Progress and challenges at the end of 2006, Tokyo.
OHCHR. (2009). *The right to adequate housing*. Fact Sheet No. 21 (Rev. 1). http://www.ohchr.org/Documents/Publications/FS21_rev_1_Housing_en.pdf. Accessed February 12, 2015.
Oliver-Smith, A. (1991). Successes and failures in post-disaster resettlement. *Disasters, 15*(1), 12–23.
Olshansky, Robert B., Johnson, Laurie A., & Topping, Kenneth C. (2006). Rebuilding Communities Following Disaster: Lessons from Kobe and Los Angeles. *Built Environment, 32*(4), 354–374(21). https://doi.org/10.2148/benv.32.4.354
Quarantelli, F. L. (1988). Disaster management crisis: A summary of research findings. *Journal of Management Studies*, *25*(4), 373–385.
Quarantelli, F. L. (2000). *Disaster Planning, emergency management, and civil protection: The historical development and current characteristics of organized efforts to prevent and respond to disasters*. Preliminary Paper #301: Disaster Research Center, University of Delaware. http://www.udel.edu/DRC/PreliminaryPapers/pp.301.pdf. Accessed March 2, 2013.

Roosli, R., Vebry, M., Mydin, A. O., & Ismail, M. (2012). Building and Planning of Post-Disaster Rehabilitation and Reconstruction. *International Journal of Academic Research*, *4*(1), 194–199.

Roseveare, C. (2013). *Rule of law and international development*. London: DFID.

Rubin, C. (1991). Recovery from disaster. In: Drabek, T. E. & Hoetmer, G. J. (Eds.), *Emergency management: Principles and practice for local government*. Washington D.C.: International City Management Association.

Rubin, Claire B., Saperstein, Martin D. & Barbee, Daniel G. (1985). Community recovery from a major natural disaster. FMHI Publications. Paper 87. http://scholarcommons.usf.edu/fmhi_pub/87

Schwab, J. (1998). Planning for post disaster recovery and reconstruction. PAS Report No. 483/484, American Planning Association, Chicago, IL.

Seleim, A., & Bontis, N. (2009). The relationship between culture and corruption: A cross-national study. *Journal of Intellectual Capital, 10*(1), 165–184. doi:10.1108/14691930910922978

Sherwood, A., et al. (2015). *Resolving post-disaster displacement: Insights from the Philippines after Typhoon Haiyan (Yolanda)*. The Brookings Institution. International Organization for Migration (IOM).

Steinberg, F. (2007). Housing reconstruction and rehabilitation in Aceh and Nias. *Indonesia-Rebuilding lives. Habitat International, 31*(1), 150–166.

Steret, S. (2013). Disaster and Sociolegal Studies. New Orleans: LA: Onati Socio-Legal Series, Quid Pro Books.

Sydow, J. et al (2004). Project-Based Organizations, Embeddedness and Repositories of Knowledge: Editorial. *Organization Studies*, *25*, 1475–1489.

Tobin, Graham A. & Ollenburger, Jane C. (1993). Natural Hazards and the Elderly. Boulder: University of Colorado, Natural Hazards Research and Applications Information Center.

Turner, B. L., et al. (2003) A framework for vulnerability analysis in sustainability science. *Proceedings of the National Academy of Sciences*, *100*(14), 8074–8079.

UNHCR (2014). Global Strategy for Resettlement and Shelter: A UNHCR Strategy 2014–2018. Available at http://www.unhcr.org/530f13aa9.pdf

Valte, M. (2000). Chapter 6. Urban Shelter. In *HUDCC 2000 Report*. Urban Partnerships Foundation. Retrieved from http://www.ombudsman.gov.ph

Walker, D., Steinfort, P., & Maqsood, T. (2014). Stakeholder voices through rich pictures. *International Journal of Managing Projects in Business, 7*(3), 342–361.

World Bank. (2013). *Building resilience: Integrating climate and disaster risk into development*. Washington DC. Available at: www.worldbank.org

World Bank. (2016). *Doing business 2016: Measuring regulatory quality and efficiency.*

Ye, Y., & Okada, N. (2002). *Integrated relief and reconstruction management following a natural disaster*. In Second Annual IIASA-DPRI Meeting, Integrated Disaster Risk Management: Megacity Vulnerability and Resilience (pp. 1–18), IIASA, Laxenburg, Austria, July 29–31, 2002.

Chapter 2
Sociology, Normative Pluralism, and Post-disaster Recovery: The Case of the Philippines

Abstract This chapter argued why the sociological normative pluralist perspective is appropriate for analyzing the unintended effects of enforcing post-disaster recovery projects in disaster-prone countries with multiple legal and social normative systems, rigid bureaucratic regulations, and high red-tape and corruption indices like the Philippines. Firstly, it traced broadly the development of disaster research in sociology and the substantive trends in theorizing hazards and disasters. Secondly, it clarified the distinction between legal pluralism and normative pluralism in socio-legal studies and explains the significance of the normative pluralist framework in evaluating the success and failure of post-disaster recovery projects. Thirdly, it highlighted the sociological significance of examining social structure in disaster events as well as the consequences of a multiplicity of actors and normative orders in post-disaster housing projects. It contended that developing countries with high legal and normative pluralism and complex social systems, such as the Philippines, are more likely to encounter significant deviations in implementing post-disaster housing projects.

Keywords Sociology of disasters · Normative pluralism · Legal pluralism · Social structure · Sociology of law · Philippines

2.1 Disaster Research in Sociology

The classical approach in disaster studies treats disasters as events that originate in the earth and atmospheric systems. It sees physical events as impinging on the built environment and on social systems with their relatively sudden onset and the casualties, damage, and disruption they cause (Tierney, 2007, p. 509). In contrast, the sociological approach treats disasters as social rather than physical events that must be defined and examined in exclusively social terms (Quarantelli, 1987, 1997). When disasters happen, sociologists rarely, if ever, view them as "natural" but rather as "the result of poorly managed interactions between environmental hazards and human society that have placed people in harm's way without adequate

V.O. Ballano, *Law, Normative Pluralism, and Post-Disaster Recovery*,
DOI 10.1007/978-981-10-5074-9_2

coping mechanisms). For sociologists, disasters are better seen as "a result of the complex interaction between a potentially damaging physical event (e.g., floods, droughts, fire, earthquakes, and storms) and the vulnerability of society, its infrastructure, economy, and environment, which are determined by human behavior" (Birkmann (ed.) 2006, p. 10). In this sense, disasters are "unnatural" phenomena (Cardona, 1993).

Sociologists generally distinguish disasters from hazards. A disaster is understood as an event in which a community undergoes severe losses to persons and/or property that the resources available within the community are severely taxed, while a hazard is a condition with the potential for harm to the community or environment (Drabek, 2004). For sociologists, the term "disaster" refers to specific events such as typhoons, hurricanes, floods, or earthquakes, whereas the term "hazard" constitutes a class of threats such as hurricanes, tornadoes, earthquakes, and so on. Hazards reflect the risk, vulnerability, or exposure confronting families, communities, or societies (Drabek, 2005, p. 4). This distinction has provided an important frame of reference for sociologists who want to use the perspectives, concepts, and methods that define the broad field of sociology (Drabek, 2005).

The sociology of disasters is said to have started to bloom with Prince's (1920) dissertation and sociological work on a technological disaster that involved a collision of two ships in the Halifax harbor on December 6, 1917 (Drabek, 2005). This was followed by natural investigations and inquiry into the conditions of panic. In the late 1940s and early 1950s, sociological research consisted mainly of a rapid response to disasters by teams from the University of Chicago's National Opinion Research Center (NORC) and the National Academy of Sciences (Fritz & Marks, 1954). The research funding during this period came mainly from the United States military organizations; thus, the research focus was on applied research that connected disaster investigations to situations in wartime and on the socio-psychological aspects of disasters.

In the 1960s and 1970s, sociological research expanded to include field studies of organizational functioning in disasters, but its focus was on the socio-psychological aspects connected with wartime and civil defense interests (Quarantelli, 1987). During this era, pioneering researchers saw disasters as contexts in which to explore organizational and collective behavior under high-stress conditions (Tierney, 2007, p. 504). The first important milestone of this approach came with the founding of the Disaster Research Center (DRC) at the Ohio State University in 1963. The DRC's founders—E.L. Quarantelli, Russel Dynes, and J. Eugene Haas—developed a research focus that examined both the organizational and emergent social behavior during and immediately after disasters (Tierney, 2007, pp. 504–505). After gathering hundreds of factual reports on the tasks and structures of groups involved in disaster situations, Quarantelli (1966) and Dynes (1970) developed an analytical tool that bears a great impact in the sociology of disasters—the well-known typology of established (regular tasks, old structures), expanding (regular tasks, new structures), extending (non-regular tasks, old structures), and emergent (non-regular tasks, new structures) organizations. This invention resulted in the shift of focus in the study of disasters from the socio-psychological to organizational aspects of disasters. The analytical device created by DRC researchers was partly responsible for the "boom years of sociological research" on disaster. Several scholars later used this important tool to generate additional hypotheses about organizational behavior in disasters (e.g., Stallings, 1978; Drabek & McEntire, 2003, p. 98).

In the 1980s and 1990s, the enormous contribution of sociology as well as social science scholars from the United States and Europe in disaster studies became significant. During this decade, the sociological perspective primarily examined the various dimensions of social and cultural difference—including race, ethnicity, age, and class—in the study of disasters (e.g., Bolin & Bolton, 1986; Perry & Mushkatel, 1986). The bulk of the sociological research during this time was directed toward topics such as disaster policy, emergency-management organizations, warning systems, and so forth and often relied on a model of system adjustment (Drabek, 1986; Quarantelli, 1996). Although more European scholars became interested in disaster research, the enormous contribution of American social-science scholars on this topic since the 1980s remained significant (van Niekerk, 2012).

One major issue confronting sociologists in studying disasters revolves around the perspectives they apply in their research. Although attempts have been made to indicate some substantive trends in the development of sociology, there is no coherent discussion in the field of sociological studies regarding the development of disaster research (Nasreen, 2004). Most sociologists do not elaborate on their theoretical perspectives that guide their field work, albeit elements of functionalism, structuralism, symbolic interactionism, and other sociological frameworks can be identified from their work. Some have pursued the insights of social constructionism and moved into research agenda that are usually ignored by those rooted within a collective-stress viewpoint. Other scholars in disaster studies are calling for a paradigm shift (Drabek, 2005, pp. 8–9).

Despite this theoretical development in disaster research, sociologists continue to neglect the normative framework in studying disasters. Also, disaster research remains underdeveloped and marginal in the field of sociology in general. Sociology of disasters remains resistant to changes in the broader sociological landscape (Tierney, 2007). And lastly, most of the disaster scholars are still American, thus limiting the research mostly to a single culture at an advanced stage of development (Laska, 1989).

2.2 Substantive Trends in Disaster Research

Understanding disasters as systemic events have always characterized sociological research on disasters from its origins during World War II to the present day, despite the fact that the current trend has clearly become multidisciplinary in more recent decades with strong intersections among the natural, social, and policy sciences (Kreps, 2001). Disaster scholars view disasters as a mix of "natural" and "human-made" or "technological" causes and as a historically conditioned process (Quarantelli, 2005). This trend can be seen in the social sciences. Over the last couple of decades, disaster sociologists, as well as scholars from psychology, political science, anthropology, geography, and disaster research centers, came to see disasters as multifaceted. Thus, current theorizing in sociology and the social sciences on disasters tend to be based on diverse orientations such as social constructionism, postmodernism, conflict-based, and political economy theories. Research on hazards and disasters during the past 30 years has been based on the notion that individuals and groups choose how to cope with or adjust to hazards in their natural and constructed environment. Consequently, public and private policies that have been developed based on this paradigm have a management strategy with the goal of decreasing hazard-related losses, organized conceptually around a four-stage cycle of preparedness, response, recovery, and mitigation (Peek & Mileti, 2002, p. 514). The existing research trend on disasters and hazards in sociology and the social sciences tends to focus on describing the conditions that can lead to disasters and achieving a prevent-and-empower disaster response. At present, three popular social concepts and perspectives, namely, social vulnerability, risk, and resilience, tend to dominate disaster research and literature in sociology as well as in the social sciences.

2.2.1 *Risk*

The first major concept and popular perspective in the current disaster research is risk. Risk research encompasses a broad range of topics on the probability of disaster events and their effects to communities. Risk scholarship flows from the work on technology; natural disasters; individual, group, and organizational decision-making; politics; stratification; environment; and social-impact assessment (Fredenburg, 1986; Dietz, Stern, & Rosa, 1993). The fields of risk assessment and risk analysis are aimed at identifying, measuring, characterizing, and evaluating the outcomes resulting from natural and technological hazards (Lowrance, 1976; Crouch & Wilson, 1982; Mitchell, 1990; National Research Council, 1993; 2006). "Disaster risk is socially distributed in ways that reflect preexisting inequalities, in that some groups are more prone to death, injury, economic loss, and psychological impairment in the wake of differing hazards (Wisner et al., 2004). Specifically, children, the elderly, women, racial minorities, the poor, persons with physical or mental disabilities, and immigrants have been identified by both disaster researchers and policy makers as especially vulnerable to the harmful impacts of disaster (Cutter, Boruff, & Shirley, 2003)."

Disaster risk is generally viewed either in a constructivist or objectivist manner. Constructivist thinking views disaster as a social construct and examines the social representations and perceptions as well as the interaction between different social actors and phenomena. It views the conditions of risk, as well as the attitudes to risk, as being rooted in societies that can inevitably lead to disasters. Objectivist or realist thinking, which is popular in the natural and physical sciences, believes that risk can be quantified and objectively judged (van Niekerk, 2012, p. 4).

Other social-science perspectives have influenced the products and risk assessments in disaster study more than sociology (Dietz et al., 1993). In particular, the field of anthropology examines how culture and ideology shape societal definitions of risk. Douglas and Wildaveshy (1982), for instance, argue that risk is not a reflection of objective reality but rather a cultural phenomenon that reflects societal and group values and that it must be interpreted in the light of their broader cultural functions. The fields of psychology and social psychology, which dominated risk research in the social sciences, focus on "how individuals perceive various risks, what factors enter into the estimation of risk, and how people make risk-related choices" (Tierney, 1999, p. 218).

Despite its popularity, not all scholars rely on the factor of risk in assessing hazards and disasters. This concept is not without criticism from sociologists. Zinn (2009), for instance, argued that it is obviously unreasonable to "over-rely" on risk because it is too narrow a concept to any sociological research. It is particularly harmful to the sociological approach, subscribing to the technical and "rationalist" understanding of risk or the economic approach grounded in instrumental rationality and a conventional model of rationality. Although some sociological work is based on rational actor approaches (Coleman, 1990; Gambetta, 1988), most sociologists find this approach unsatisfactory in dealing with situations in which others are involved. Although research using risk framework recognizes that risks are shaped by structural and institutional contexts, it has no integrating theory. The distinctive contribution of this perspective is its emphasis on the role of shared ideas and normative frameworks, as understood in terms of the contribution of cultural and social factors, to the understanding and prioritizing of risks and responses to them among all those involved (Zinn & Taylor-Gooby, 2006).

2.2.2 *Social Vulnerability*

Another popular concept and theorizing in the sociology of disasters is social vulnerability. The social-vulnerability perspective in viewing disaster has grown in prominence in the sociology of disaster literature during the past decade. Although this concept is often cited by disciplines of applied research, there is really no common working definition for the term (Birkmann & Wisner, 2006). The various definitions of vulnerability in the literature depend on the conceptual models and frameworks of authors (Adger, 1999; Downing, Olsthoorn, & Toi, 1999; Cutter, 1996; See & Porio, 2015). "Vulnerability is a term used in the field of risk, hazard, and disaster management as well as in the areas of global change and environment and development studies" (Weichselgartner, 2001). The UNISDR (2004, p. 16) defines it as "the conditions determined by physical, social, economic and environmental factors or processes which increase the susceptibility of a community to the impacts of hazards" (Palliyaguru et al., 2014). It can also refer to "the susceptibility of groups to the impacts of hazards, as well as their resiliency, or ability to adequately recover from them" (Cutter & Emrich, 2006, p. 103). The concept of social vulnerability "highlights differences in the human capacity to prepare for, respond to, and recover from disasters. It varies over space and time, and among and between social groups, largely due to differences in socioeconomic and demographic characteristics" (Hummel, Cutter, & Emrich, 2016). Scholars who employ this concept believe that vulnerability is partially the product of social inequalities, i.e., those social factors that influence or shape the susceptibility of various groups to harm and that also govern their ability to respond (Cutter, Boruff, & Shirley, 2003, p. 243). Disasters, such as volcanic eruption, flood, or typhoon, are only a triggering agent. What determines vulnerability is the quality of the existing infrastructure or degree of resilience of the affected community to the triggering agent. To them, the vulnerability can never be isolated from the social, cultural, economic, and political realms that influence decisions regarding settlement patterns and methods for building (McEntire, 2001). It is a product of decisions that are structured by the normative systems, be they legal or cultural. These decisions create the vulnerable environment. Thus, vulnerability is ultimately a product of an interplay of various normative systems of individuals and groups in society that created a disaster-prone situation.

The concept of vulnerability in research derives from a specific set of historical circumstances relating to abuses in biomedical research. One major criticism of this approach is that scholars who use this concept have labeled so many people and groups as vulnerable that the concept has lost much of its force. The vulnerability paradigm in disaster research has limitations. The concept of vulnerability is an extraordinarily elastic concept, capable of being stretched to cover almost any person, group, or situation, and then of being snapped back to describe a narrow range of characteristics such as age or incarceration. The concept gathers people with widely varying characteristics and capacities under one large umbrella without analyzing whether or not they all fit. It is both too broad because it includes too many categories and too narrow because it fails to include others at risk for different reasons other than their cognitive or dependent status (Levine, 2004, p. 398). In disaster research, participants should not be automatically considered vulnerable unless they are legally designated as such.

2.2.3 Resilience

Resilience is another popular perspective in disaster research. The term "resilience" has its roots in the sciences of physics and mathematics. It was originally used to describe the capacity of a material or system to return to equilibrium after a displacement (Bodin & Wiman, 2004; Gordon, 1978). When applied to people and their environments, "resilience" is fundamentally a metaphor that implies strength, capacity, elasticity, and evolution (Alexander, 2013a). Although this concept began to be employed as the inverse of human vulnerability in the disaster risk–management community in the late 1970s with increased use in recent years (Birkmann, 2006; Gaillard, 2010; Torry, 1979), some scholars believe that resilience is not to be seen as the opposite of vulnerability. They accede that both concepts may overlap, but they assert that it can best be understood as discrete concepts because perceiving these as "two sides of the same coin" leads to unproductive circular reasoning (so that no new policy options arise despite the linguistic turn) (Klein, Nicholls, & Thomalla, 2003; Matyas & Pelling, 2015).

The term "resilience" has different names. "Ecologists call it an adaptation, economists call it is coping capacity, anthropologists call it bounce back better, and in engineering, it is best known as the capacity of a structure to withstand shock while retaining function" (Dahlberg et al., 2015, p. 44). The original notion of the term comes from a Latin word that means "jump back" or "bounce back" (Manyena et al., 2011). In the disaster context, resilience can be understood as the ability of people to recover within the shortest possible time with minimal or no assistance (Malalgoda, Amaratunga, & Haigh, 2014). Although many interpretations of the term abound in the disasters community, this understanding of resilience has proved to be persistent." Thus, the United Nations Strategy for Disaster Reduction (2009) defines resilience as:

> the ability of a system, community or society exposed to hazards to resist, absorb, accommodate to and recover from the effects of a hazard in a timely and efficient manner, including through the preservation and restoration of its essential basic structures and functions.

The rapid increase of the concept of resilience in the disaster risk–management (DRM) policy landscape started with the Hyogo Framework for Action 2005–2015 (HFA), whose subtitle, "Building the Resilience of Nations and Communities to Disasters," places the concept at the core of DRM aspirations (Matyas & Pelling, 2015a, b). "The intimate connections between disaster recovery and the resilience of affected communities have become common features of disaster risk reduction programs since the adoption of the HFA. Thus many disaster research is now giving more attention to the capacity of disaster-affected communities to 'bounce back' or to recover with little or no external assistance following a disaster. This highlights the need for a change in the disaster risk reduction work culture, with the stronger emphasis being put on resilience rather than just need or vulnerability" (Manyena, 2006).

Resilience as a concept and perspective to view disasters has the potential to offer a more systemic and cross-cutting approach to disaster-risk reduction, climate-change adaptation, and the humanitarian sector. Resilience is an important goal for two reasons. First, because the vulnerability of technological and social systems cannot be predicted completely, resilience—the ability to accommodate change gracefully and without catastrophic failure—is critical in times of disaster (Foster, 1997). If people know exactly when, where, and how disasters would occur in the future, they could engineer their systems to resist them. Because hazard planners must cope with uncertainty, it is necessary to design cities that can cope effectively with contingencies.

The concept of resilience has since been applied to describe the adaptive capacities of individuals (e.g., Bonanno, 2004), human communities (e.g., Brown & Kulig, 1996/97; Sonn & Fisher, 1998), and larger societies (e.g., Adger, 2000). Because references to resilience have continued to increase, so too have criticisms that the concept may be inappropriate, imprecise, or "glittery" (Norris et al., 2007). Some skeptics argue that the pursuit of resilience, particularly community resilience, is laudable but impractical. Using a conceptual framework based on theoretical models of mitigation, recovery, and structural–cognitive interaction, Tobin (1999), for instance, examined data from the state of Florida to assess the possibility of developing sustainable, resilient communities. Analyzing the state as a whole rather than individual cities, he concluded that major changes in political awareness and motivation would be necessary to overcome obstacles to resiliency and sustainability from Florida's existing demographic traits, spatial patterns, and hazard conditions (Godschalk, 2003). Resilience, like social vulnerability, and risk as concepts and perspectives in disaster research may characterize some social aspects of disasters but not their entire structural components. The key to a holistic understanding of the nature of hazards and disasters requires a sociological examination of the stable pattern of social relationships existing in a particular group society or group where the disaster event took place or what sociologists call "the social structure."

2.3 Sociology of Disasters and the Social Structure

The study of the social structure and social system of disaster situations has apparently been sidelined in disaster research. Yet, disasters do not only alter the geophysical components of society but also its social structure. After a disaster event, the existing social structure of society is disrupted. Its stability is shaken, and the prevailing social system is replaced by a new emergent structure that tends to be incoherent, polycentric, and pluralist in the normative system. Sociologists do not just study the natural or human-made "disaster agents," such as hurricanes, flood waters, or hazard spills, but also the disruption of everyday social activities resulting from the potential or actual impact of these agents (Fischer, 2003, p. 93). Some sociologists argue that disasters may expose the key values and structures that define communities and the societies they comprise (Drabek, 2005, p. 3). In this case, they discover that the study of structural change occurring during and after the precipitous event is crucial in the study of disasters. Thus, they examine the altered social structure in the aftermath of the impact of a disaster agent when it causes widespread destruction and distress. In short, disaster sociologists study the change in social structure under specific circumstances (Fischer, 2003, p. 94).

The focus of the sociological research must be the status quo (i.e., the structure before the disaster) and the disruption (or big mess) thereafter leading to the adjusted social structure during the post-disaster recovery phase. The greater the degree of disruption (in terms of scale, scope, and time), the more the social structure is affected, thus leading to the emergence of new norms and roles replacing the routine with the "more appropriate" behavior, e.g., search and rescue, feeding survivors (Fischer, 2003, p. 97). In this case, disaster-governance regimes become polycentric, multi-scale, and weak in integration as well as nested within and influenced by overarching societal-governance systems (Tierney, 2012). Therefore, coordination between state and the various post-disaster actors and organizations—each with their own normative systems, especially in cases of large-scale disasters—becomes problematic. The greater the impact of the disaster, the greater is the number of actors and organizations that are involved in the recovery process, thus making the normative order become more heterogeneous and intricate. Then the implementation of post-disaster recovery (PDR) projects becomes prone to more negative unintended consequences. The enforcement of the intended PDR goals becomes amorphous and polycentric which can be detrimental to the rehabilitation of disaster victims.

2.4 The Social System and the Normative System

Studying the altered social structure during the post-disaster situation requires examining the normative orders. The social system is a system of action, and its structural aspects are the relatively stable interactions of individuals around common norms. It is made up of the interactions of individuals governed by norms. The regularity or patterning of interaction becomes possible through the existence of norms, which control the behavior of the actors. Because individuals share the same "definition of the situation" in terms of such norms, their behavior can be intermeshed to produce a "social structure." Indeed, a stabilized social system is one in which behavior is regulated in this way and, as such, is a major point of reference for the sociological analysis of the dynamics of social systems (Lockwood, 1956, p. 135). The actual disaster event can initially trigger solidarity among victims and rescuers, but a post-disaster setting can evoke pluralist and discordant responses from various actors and groups during the reconstruction process. The regularity of the social system is gradually replaced by a novel one, made of a loose amalgamation of normative orders with a weak overall monitoring and law-enforcement system to guide post-disaster recovery.

With the plurality of actors during and after disasters, perceptions can vary on how to deal with the disaster event. "Although there has been rising global awareness and lesson sharing on the human catastrophes associated with the proliferation of natural disasters, there remains many misunderstandings on appropriate approaches to recovery, given the specific socio-cultural, political and economic contexts of the affected zones" (Brazzard & Raffin, 2011, p. 417). Bankoff and Hilhorst (2009) aptly describe the diversity of views or ways of "seeing" disasters by different actors and organizations:

> Different actors 'see' disasters as different types of events and, because they perceive them as such, they prepare for, manage and record them in very different ways. State and state agencies 'see' disasters in one way; the people directly affected 'see' them in another (not least according to the nature and extent of their vulnerabilities); non-governmental organizations (NGOs) involved in providing needed services to communities 'see' them in still other ways (depending on their ideological complexion); whereas scientists, technocrats and experts have their own variants of 'seeing'. The media, too, if it does not exactly 'see' risk in a particular way certainly helps to shape public discourse about it (Benthall, 1993; Stallings, 1990, p. 82). These differing 'seeings' not only shape how actors and stakeholders perceive disasters but also largely influence what actions are taken before, during and after an event. Actors and stakeholders do not respond directly to situations: rather, the purpose and nature of their actions are determined through mediating orientations that dispose them to act in certain ways (p. 687).

Varying worldviews and perceptions imply different actions in responding to the post-disaster process. In any social structure, the actions of actors are shaped by their normative orientation. In a post-disaster setting, the constellation of legal and social normative orders tends to overlap and expand depending on the number of actors joining in the rehabilitation and reconstruction process. "Sources of governance in disaster management which are traditionally associated with one set of actors (private governance with private actors or international law with international institutions, for example) are increasingly co-produced through the involvement of a range of actors in their formulation and implementation for reasons of capacity, resources, or reach" (Newell, Pattberg, & Schoeder, 2012, p. 366). With the rise of megacities, this growing multiplicity of actors and sources of governance in disaster management requires that disaster sociologists examine the plurality of normative systems and their influence on the implementation of projects in the post-disaster recovery process.

2.5 The Normative Pluralist Approach

The substantial increase in disasters around the world has led to an increase of international, national, and local laws and norms on disaster management. Despite the emergence of international norms, comparatively little scholarly attention has been given to compliance of disaster efforts with these normative frameworks as well as on how international laws and norms interact with national and local norms.

There is apparently an absence of scholarly attention on the legal and normative frameworks that underpin disaster situations (McDermott, 2013, p. 1). Although there is a growing body of research on disaster governance, examining the intertwining legal and non-legal normative systems after large-scale disasters and how they affect post-disaster projects, such housing and resettlement, is generally missing in disaster research and particularly in sociology of disasters.

Sociology has explored some societal aspects of disasters. However, it has not scrutinized closely the legal and social normative orders that shape the PDR in disaster-prone countries with rigid regulatory systems and collectivist cultures. Sociology has examined the behavioral, organizational, and institutional aspects of natural disasters but not the normative systems that influence disaster management, particularly during the PDR period. Consequently, the sociological stance of the subfield of the sociology of law, which specifically examines the dynamics of law and the social order, continues to be ignored by disaster scholars. If the sociological perspective sees disasters as social phenomena, then sociologists must analyze the normative structures that frame, interpret, and actualize disaster response and recovery. Together with other legal scholars, it is incumbent upon sociologists to undertake an intensive effort to examine and reform the interlocking structures of governance and regulation that pertain to disasters (Farber, 2012, p. 20).

Seeing disasters as systems rather than arenas of normative conflicts has sidelined the socio-legal approach in sociological research. The mainstream sociological research, which is guided either implicitly or explicitly by systems concepts, views disasters as enhancing social solidarity and suppressing conflict (Tierney, 2007, p. 506). The pioneering work of sociologists Quarantelli, Dynes, and White has established the parameters of mainstream disaster research that views disaster situations toward atomism rather than pluralism; thus, it unwittingly disregards the normative pluralism aspects of disaster situations (Iversen & Armstrong, 2008). The fields of social-legal studies, including criminal justice and the sociology of law, which primarily examine the normative orders, have been relatively absent or, at least, have not been explored well in the sociology of disasters (Deflem, 2012, p. ix). This neglect seems to be ironic because disaster response is complex and governed by a multiplicity of actors and normative systems. A post-disaster situation can consist of multiple competing formal and informal normative forces instead of a unified and interrelated system that suppresses conflict. The emergent character of a post-disaster recovery situation requires an ad hoc multi-organizational set-up, which can lead to discordant and conflicting responses from the various disaster agencies acting according to their normative and regulatory frameworks and political orientations.

The normative pluralist approach views quantity and harmony of different formal and informal regulations that underpin disaster management as crucial in effecting successful post-disaster recovery projects. The more actors and normative systems surrounding a PDR project, the more difficult it is to achieve the intended goals. In this case, the success or failure of regulatory compliance of the PDR project will depend to a large extent on the readiness of actors to accept responsibility to implement it at every level (Roosli & O'Keefe, 2011). PDR projects require a concerted effort that will support the foundations of community sustainability and capacity building, which will eventually decrease risks and vulnerabilities to future disasters (Burton & Kates, 1993). Cooperation from all levels in the disaster mechanism and the public is the ultimate goal to deliver a national disaster program (Roosli & O'Keefe, 2011).

Large-scale post-disaster recovery project is essentially multiple and pluralist in terms of normative and regulatory orders. The government, which often leads the entire process, is assisted by a variety of stakeholders and organizations, both public and private as well as local and international. These stakeholders involved in PDR projects, such as housing and resettlement projects, possess their own set of formal and informal norms of "doing things" that determine their "repertoire of disaster-recovery actions." Their norms tend to clash or conflict with one another. In post-disaster housing projects, it is a near-impossible balancing act to satisfy all parties involved (Daly & Brassard & Raffin 2011). Thus, coordinating the needs and aspirations of beneficiaries with the means and systems of donors and implementers can be complex, highly context specific, and rarely seamless (Fallahi, 2007; Tas, Cosgun, & Tas, 2007; Saunders, 2004; Daly & Brassard, 2011). "Cooperation from all levels in the disaster mechanism and the public is the ultimate goal in order to deliver a national disaster program" (Roosli & O'Keefe, 2011). However, there is usually a lack of unified and coherent legal framework and law enforcement system that can monitor, coordinate, and harmonize the various normative orders of actors and organization to successfully complete the PDR projects. As a result, the accountability of PDR actors becomes difficult to ascertain, and corruption of various forms can thus penetrate the different phases of the project's implementation. With the lack of a stable law-enforcement system that effectively regulates people's action in large-scale calamities, disaster behavior that implements post-disaster projects can easily be justified by actors in a variety of grounds—legal, political, economic, social or cultural—other than what the official rules dictate.

Different types of organizations engaged in PDRs are expected to share tasks and resources. But there is usually no clear-cut responsibility or familiarity due to the crossing of jurisdictional boundaries and the absence of standardization during emergency situations (Auf der Heide, 1989). With the usual absence of a clear standard of measuring accountability during a post-disaster situation, actors and organizations in the PDR system can engage in a power struggle to influence the outcome of the different aspects of the PDR projects. In one disaster-recovery project, for instance, actors and groups can become "insiders" and "outsiders," each extending their disaster response according to their goals and normative frameworks. A study by Espia and Fernanadez (2015) on government and NGO response to the Guimaras oil spill in the Philippines in 2014, for instance, revealed that in the absence of formal rules, the NGOs tended to counteract and compete with the local government's designed response framework, thus reaffirming their outsider status in the rehabilitation process. The so-called entrepreneurial brokers can also emerge in a post-disaster setting and play a major role in governance networks of the PDR projects. These brokers can play as "middlemen" or intermediaries of the various network ties to mobilize the coordination system during post-disaster recovery

(Saban, 2015). The "corrupt" and illegal practices by some international and local disaster groups that were allowed or were contracted by the government in housing and resettlement projects in the spirit of what Klein (2008) calls "disaster capitalism" indicate that PDR is an arena of contestations and consisting of conflicting normative forces.

2.6 Legal Pluralism and the Social Normative System

"Scholarly assumptions about the state as a unitary actor and as a homogeneous, monolithic unit and about state centricity as the foundation of international politics have gradually crumbled in favor of more multicentric views of the world" (Newell, Pattberg, & Schoeder, 2012, pp. 368–369). The monist approach to the state's legal system has been debunked by legal pluralism. The idea of "legal pluralism" emerged in the early 1970s as a counterbalance to the the-dominant notion of "legal centralism," according to which law is and should be the law of the state, uniform for all persons, exclusive of all other law, and administered by a single set of state institutions. In lieu of such state-centric and monolithic conceptions of law, advocates of legal pluralism claimed that law is not a single system necessarily linked to the state as a unified entity, but rather a complex of overlapping systems or normative orders" (Sartori & Shahar, 2012, p. 638). The state's legal system determines whether governance is multiple and does not operate in a unified manner.

The normative pluralist approach does not only see a pluralism of laws but also of informal norms in the social order. It sees legal pluralism as one of the species of the greater normative pluralism in society. It views the social structure as one system of interacting, conflicting, and intertwining legal and social normative systems. The social realm consists of a plurality of interacting, overlapping, active regulatory systems of every kind from religious systems, to corporations, to sports leagues, to the family. The normative-system relationships between the official legal system and the social normative system are "various and complex, sometimes conflicting, sometimes complementary, sometimes benign, sometimes cautious engagement, sometimes ships passing in the night. People can invoke these coexisting systems in various ways for various instrumental and normative reasons, regularly and deliberately pitting one system against another. The official legal realm affects and is affected by the multiple regulatory orders in the social realm; the regulatory orders in the social realm affect and are affected by the official legal realm" (Tamanaha, 2008, p. 90).

When applied to disaster management, normative pluralism views disaster behavior of various actors and groups in response to a massive disaster as emanating from different normative or regulatory regimes, some of which may be formed after the disaster (Schneider, 1992). Thus, significant incoherence and incongruity of interpretation and action in implementing PDR projects can lead to more negative unintended consequences and deviations that can ultimately affect the success and failure of PDR projects. Outside the official legal system is the social realm of intercourse and regulation. This domain consists of socially created and coordinated regulatory systems produced by individuals acting on shared ideas and beliefs with particular projects in specific contexts. They have effects, consequences, and implications of all kinds for official state legal systems as well as for the social arena generally. In disaster settings, a multiplicity of actors and governance regimes operate in diverse and overlapping spheres of authority and normative systems (Newell, Pattberg, & Schoeder, 2012, pp. 368–369).

As society becomes more urbanized, disaster effects become more large scale. With the growing magnitude of disaster effects, disaster response becomes more complex, which requires a large-scale infrastructure beyond the capacities of victims to mobilize themselves or within the community. Consequently, disaster response, which used to be simple and limited to small territories, has become massive and complicated due to environmental destruction caused by intensifying urbanization and industrialization, particularly in Asia (Douglass, 2013). The network of actors and organizations involved in rehabilitation becomes complex and includes (1) public actors such government and civil society organizations, (2) international and regional environmental-governance arrangements, (3) public–private partnerships such as alliances between the government and NGOs, transgovernmental and transnational networks, and partnerships, and (4) private actors (Newell, Pattberg, & Schoeder, 2012). Each kind of actors can have its own set of social networks and normative orientations set in a multi-level arrangement.

A disaster response within a more cohesive legal order can more likely achieve its intended goals compared with a disaster situation where the legal orders are complex or full of gaps with the absence of a strong and centralized post-disaster law-enforcement system. In the latter case, the social normative orders, with their informal regulation and enforcement systems, are more likely to create negative unintended consequences that can shape the final response of PDR projects. Public and private organizations—as well as informal groups, power cliques, foreign and local donors, NGOs, POs, and other groups—tend to have their own established formal and informal rules of engagement within their own PDR domains. When they participate in common PDR projects, their normative orientations often clash with each other. This conflict can also be seen among public organizations. Different government agencies must operate at cross-purposes to achieve the overall PDR goals; conflict and inconsistencies within and between relevant agencies can occur, which eventually hamper law enforcement.

2.7 Regulation in High-Context Cultures and PDR Housing

The regulatory normative system, more specifically the non-legal cultural regulatory system, can be crucial to the success of PDR projects in high-context cultures. Sociologists and anthropologists have started to examine the latent and informal communication and other cultural dimensions of contracting. Edward T. Hall distinguishes low-context from high-context cultures. In high-context cultures, many things are left unsaid and require cultural context to illuminate the meaning. North America and Western Europe are regarded as low-context cultures, which value logic, facts, and directness, whereas high context cultures in Asia, Middle East, Africa, and South America are relational, collectivist, intuitive, and contemplative emphasizing interpersonal relationships (Hall, 1976). Hooker (2003) divides cultures into rule-based and relationship-based and assumes a continuum between different cultures that results from multiple combinations of features of both types of cultures (Hooker in Nystén-Haarala, Bogdanova, Kondakov, & Makarova, 2015, pp. 113–114). The Global North, with its low-context cultures, follows predominantly a rule-based normative order, thus putting more emphasis on legal and formal regulation in pursuing PDR projects. But the Global South, with its high-context cultures, primarily implements a relational and personalistic approach to contractual relations in PDR projects.

The existence of multiple relationship–based normative orders vis-à-vis the state's legal system in high-context cultures, such those in Southeast Asian countries, poses serious problems in regulating and implementing PDR housing projects. Regulation determines access and use of resources for PDR projects. This can be formal or informal as well as legal or social. The formal and legal regulations are provided by the state and its bureaucratic apparatuses. However, the informal and nonlegal regulations that intertwine with legal regulation can be determined by the social and cultural normative standards of social groups in society. The official legal realm affects and is affected by the multiple regulatory orders in the social realm; the regulatory orders in the social realm affect and are affected by the official legal realm (Tamanaha, 2008). The official legal system interacts with the social realm of intercourse and regulation. And this social realm is full of a plurality of interacting, overlapping, active regulatory systems of every kind from religious systems, to corporations, to sports leagues, to the family. "Like the official legal system, these are socially created and coordinated regulatory systems produced by individuals acting upon shared ideas and beliefs, with particular projects in specific contexts. They have effects and consequences and implications of all kinds for official state legal systems, as well as for the social arena generally" (Tamanaha, 2008). When formal regulatory mechanisms are weak or ineffective, communities seek other informal regulations to translate their preferences into reality. When formal regulation leaves a considerable gap between the actual performance and what people locally prefer, the informal normative orders, or what some scholars call "practical norms," take the stage, so to speak, to informally regulate areas that are not legally regulated. This influence of social-regulatory systems can breed many negative unintended effects when implementing PDR projects.

In an ineffective formal regulation regarding the access of disaster victims to government housing, informal regulations of various agencies involved in the housing project can control such access. This informal regulation that governs access to a public housing project for disaster victims is determined by heterogenous informal normative orders. Therefore, the regulation that governs the availability of post-disaster housing to beneficiaries is an interplay of formal and informal regulations that are highly unfavorable to the most vulnerable group of disaster victims.

2.8 Red Tape, Corruption, and PDR in Southeast Asia

Aside from the informal and social normative systems, red tape and rigid bureaucratic regulation can also be crucial for the success or failure of PDR projects in disaster-prone countries of Southeast Asia. Some countries with high juridical legal pluralism in Southeast Asian countries are also countries with high red-tape and corruption indices. Indonesia and the Philippines, for instance, are frequently ranked high in corruption index by Transparency International (TI) and listed low by the yearly surveys of the World Bank on the ease of the doing-business due to bureaucratic burdens and rigid business regulation. Interestingly, these countries are disaster-prone and have serious problems in post-disaster recovery programs, especially concerning the construction of resettlement and emergency-shelter projects. Cultural and "practical" relational norms tend to sidestep the rule-based ideals of government-initiated PDR housing projects.

The normative landscape in the Global South, particularly in the collectivist cultures of countries under Association of Southeast Asian Nations (ASEAN) countries, tend to be more personalistic, informal, in–group oriented, and pluralist in normative systems compared with the individualist and rights-based orientation of contracting in the Global North. This pluralist legal and cultural normative system can breed red tape and corruption, which can deviate PDR goals in large-scale reconstruction projects. The term "red tape" refers to "the rules, regulations, and procedures that entail a compliance burden but do not achieve the functional objectives of the organization" (Bozeman, 1993). To minimize red tape in an environment of legal pluralism, there must be a high degree of trust between rule-producers and subjects of the rule. In some cases, those producing new and extensive rules and regulations are in a dynamic relationship with those complying with the rules and regulations, such that the compliance level may depend on particular specifications of the rules and more on the degree of trust and the recent history between producers of rules and those affected by them (Dirks & Ferrin, 2001). In many cases of post-disaster recovery projects in developing countries such as the Philippines, the rules and regulations for low-cost housing and relocation are already in place, but they do not necessary address the PDR concerns of disaster victims. Thus, the disaster victims must comply with numerous pre-existing and unnecessary legal and bureaucratic requirements that delay their recovery and resettlement. In this case, these obligations and procedures constitute red tape and additional burdens to disaster victims. To avoid red tape and create short-cuts in these tedious legal and bureaucratic procedures and to expedite the PDR resettlement process, corruption occurs. In a high-context and collective culture where legal pluralism and red tape dominate government regulatory system, the informal and cultural norms can create procedural short-cuts colonize the rule-based PDR process of the post-disaster housing projects of the state.

2.9 The Philippine Case

It has been observed that the dominant response and action on disaster management in Southeast Asia had been on post-disaster activities and particularly on emergency response" (Bildan, 2003; Jegillos, 2003). Of the 10 countries in the world most imperiled by climate change (in terms of the number of people likely to be affected), 4 are in Southeast Asia: Vietnam, Indonesia, Thailand, and the Philippines (Elliot, 2012, p. 40). The Philippines is one of the most disaster-prone Southeast Asian countries. Approximately 70% of its disasters are due to hydro-meteorological phenomena such as typhoons and flooding. Typhoon Ketsana, which hit the Philippines in 2009, showed differential vulnerabilities of the population. The most vulnerable regions and communities in the country are those that are highly exposed to the changes expected in the climate and have limited adaptive capacity. The Philippines has limited economic resources, low levels of technology, poor information and skills, poor infrastructure, unstable or weak institutions, and inequitable empowerment and access to resources have little capacity to adapt and thus is highly vulnerable to disasters (IPCC, 2001) (Penalba, Elazegui, Pulhin, & Cruz, 2012, p. 310). Disaster victims in the Philippines are usually the urban poor who are totally dependent on social services. These people are usually less able to respond effectively to disasters (Cutter, Boruff, & Shirley, 2003; Morrow, 1999; See & Porio, 2015). The socially and economically marginalized population are mostly ignored during disaster recovery.

Aside from being vulnerable to climate change and disasters, the Philippines is also known for legal pluralism and bureaucratic burdens in real-estate business. Establishing a post-disaster housing and relocation program for disaster victims, such the flood victims of Typhoon Ketsana, requires compliance with the multiple legal and bureaucratic requirements. The rigid regulatory requirements attending housing and land development contribute significantly to greater costs and increased inaccessibility of low-cost housing. Due to the different permits, clearances, licenses, and procedures that must be obtained from various agencies other than that of housing, completion of these requirements often takes 2.5–3 years (Valte, 2002). Red tape and corruption often dominate the low-cost housing system that assists homeless disaster victims in the country. To circumvent the tedious process of securing permits and registrations, real-estate developers and contractors often provide bribe money to some unscrupulous government bureaucrats to speed up real-estate transactions. This "hidden cost" of corruption is ultimately passed on to end-users in the form of high-priced and/or sub-standard housing products" (Valte, 2002). To government-accredited contractors and developers, post-disaster housing, just like any other low-cost housing in the government, is a huge business enterprise. The extra costs they incur in securing the necessary permits, licenses, and other bureaucratic requirements due to corruption are passed on to the government financing institutions, which are obliged under a joint venture agreement to purchase their housing units. Ultimately, the homeless typhoon victims will eventually become recipients of the developer's poorly constructed houses in the government's disaster-prone relocation sites.

Finally, access of the economically disenfranchised disaster victims to housing projects usually requires political patronage or sponsorship of politicians. The rigid formal regulation that needs multiple requirements for housing applications implies more suffering for disaster victims. It requires sponsorship from politicians and government officials from different public agencies who can facilitate the approval of their housing applications. The informal regulation based on cultural norms further increases the difficulty of the disaster victims to own their own houses in relocation areas. In this case, the popular informal normative systems, which are based on some dominant Filipino values —also known in Philippine politics such as *palakasan (patronage), utang-na-loob (debt-of-gratitude), pakikisama (comradeship),* and the *Padrino (patronage)* system. These cultural values tend to dominate the official normative system to overcome red tape, rigidity in housing regulation, and bureaucratic corruption. This dominance of informal cultural norms can eventually lead to negative unintended effects to resettlement and housing projects and sidelining of the government's rule-based PDR goals as articulated in the country's primary disaster-management legislation, i.e., the Philippine Disaster Risk Reduction Management Act of 2010 (PDRRMA).

2.9.1 Super Typhoons and Damaged Houses in the Philippines

Owing to a greater demand in the market, the backlog for Philippine housing currently stands at 5.5 million. Industry players said that this might even increase through 2030 if the demand for socialized houses or mass houses is not addressed by the government and private sector. Socialized housing under the Urban Development and Housing Act of 1992 (UDHA) involves houses and lots or home lots only developed by the government or private sector for underprivileged and homeless citizens.[1]

One of the major cause of this backlog is the partial and total destruction of houses, especially those of the urban poor, by destructive typhoons and other forms of disasters that usually hit the country every year. The Philippines is the third most disaster-prone country in the world with 50.3% of its total 300,000 km^2 area (across 7107 islands) susceptible to disasters. Approximately 81% of the population is vulnerable to hazards and disaster risks ranging from landslides, storms (and typhoons), floods, storm surges, earthquakes, tsunamis, droughts, and volcano eruptions. During the last decade, the country has been hit by major disasters that rendered thousands of families homeless (Environmental Science for Social Change, 2014, p. 6). In all of these super typhoons, the poor were usually the most

---

[1]"Philippine housing backlog is 5.5 M SHDA targets to build a million units by 2016". Retrieved from http://www.philstar.com/cebu-business/2015/07/13/1476445/philippine-housing-backlog-5.5m-shda-targets-build-million-units.

affected group and rendered homeless by the disaster event. In urban areas, the poor who concentrate in informal settlements at high-risk areas, such as floodplains, were the most vulnerable to flooding and home damage. Similarly, in rural areas, it is the poorest who end up living in dangerous areas, such as river embankments, that received most of the effects of these super typhoons (World Bank, 2009).

Typhoon Ketsana is the first super typhoon in recent memory that strongly hit the Philippines and destroyed thousands of houses (185,004). Tropical Storm Ketsana, a category I storm, (locally known as Typhoon Ondoy) hit the Philippines on September 26, 2009. It brought an unusually high volume of rain, which inundated the central part of Luzon. During the 12-hour period of the storm on September 26, the rainfall was recorded as approximately 450 mm, considered an extremely rare occurrence. Thus, it caused extensive flooding in the Metro Manila area and the neighboring Rizal province including the cities of Antipolo, Makati, Malabon, Marikina, Muntinlupa, Pasig, Quezon, San Juan, Taguig, and Valenzuela (World Bank, 2009, p. ix). The National Disaster Risk Reduction Management Council (NDRRMC), the government's national-disaster agency, estimated the total damage of Ketsana at PHP 10.9 billion. Most of the homeless victims were assigned to different government resettlement areas.

Ketsana was not the only super typhoon that hit the Philippines and damaged thousands of homes in the current decade. As listed in Table 2.1, six more super typhoons followed Ketsana after September 2009: Parma (Pepeng), Megi (Juan) Nesat (Pedring), Washi (Sendong), Bopha (Pablo), and Haiyan (Yolanda). Tropical Storm Ketsana was quickly followed by another super typhoon named Tropical Storm Parma (locally known as Pepeng), a category III storm that affected the Philippines during October 3–9, 2009, following an irregular path that crossed over Central and Northern Luzon three times. It initially brought powerful winds with gusts of 230 km/h in an extended period of heavy rains with cumulative rainfall amounts exceeding 1000 mm in some areas (World Bank, 2009). As listed in Table 2.1, super typhoon Parma resulted in an estimated total damage of PHP 27.2 billion and 61, 869 destroyed homes.

Super Typhoon Megi followed Ketsana and struck the Philippines from October 16–18, 2010. Megi (locally known as Juan) made landfall in the northern provinces, namely, Cagayan, Isabela, Kalinga, Mt. Province, and Ifugao. As Megi continued to pummel on these provinces with strong winds and rain, roads and bridges were damaged and left impassable. Families were relocated to safer grounds for fear of landslides and flooding. Megi carried maximum sustained winds of 225 kph.[2] The NDRRMC estimated the total damage of Megi at PHP 12.0 billion, which is a lower figure compared with Ketsana and Parma. However, the number of damaged homes was 148,222, which is a much greater figure than that for Parma (61,869) and a little lower compared with that for Ketsana (185,004).

[2]"Typhoon Megi hits northern Philippines: Assessment in Cagayan and Isabela underway." Retrieved http://www.worldvision.org.ph/news/typhoon-megi-hits-northern-philippines-assessment-cagayan-and-isabela-underway.

Table 2.1 Super typhoons from 2009 to 2016 and their estimated total number of partially and completely damaged houses

Name	Date	Total loss	Damaged houses		
			Totally	Partially	Total number
Ketsana (Ondoy)	September 26, 2009	PHP 10.9 Billion	30,082	154,922	185,004[a]
Parma (Pepeng)	October 3–9, 2009	PHP 27.2 Billion	6807	55,062	61,869[b]
Megi (Juan)	October 16–18, 2010	PHP 12.0 Billion	30,048	118,174	148,222[c]
Nesat (Pedring)	September 17, 2011	PHP 15.4 Billion	6852	42,638	49,490[d]
Washi (Sendong)	December 15, 2011	PHP2.0 Billion	13,585	37,559	51,144[e]
Bopha (Pablo)	December 4, 2012	PHP 43. 2 Billion	89,666	127,151	216,817[f]
Haiyan (Yolanda)	November 7–9, 2013	PHP 93.0 Billion	518,878	493,912	1,012,790[g]
Haima (Lawin)	October 19, 2016	PHP 3.7 Billion	13,964	76,071	90,035[h]

[a]NDRRMC Update 24–27 September 2009. Retrieved from http://www.ndrrmc.gov.ph/attachments/article/1543/Update_Final_Report_TS_Ondoy_and_Pepeng_24-27SEP2009and30SEP-20OCT2009.pdf
[b]NDRRMC Update 24-27 September 2009. Retrieved from http://www.ndrrmc.gov.ph/attachments/article/1543/Update_Final_Report_TS_Ondoy_and_Pepeng_24-27SEP2009and30SEP-20OCT2009.pdf
[c]NDRRMC Update 30 October 2010. Retrieved from http://www.ndrrmc.gov.ph/attachments/article/1554/SitRep_No_28_Typhoon_Juan_Issued_On_30OCT2010.pdf
[d]NDRRMC Update. Retrieved from http://reliefweb.int/sites/reliefweb.int/files/resources/NDRRMC%20Update%20SitRep%20No%2021%20re%20Effects%20of%20TY%20PEDRING%205%20Oct%202011%206%20AM.pdf
[e]NDRRMC Update 25 October 2016. Retrieved from http://www.ndrrmc.gov.ph/attachments/article/1347/Final_Report_on_the_Effects_and_Emergency_Management_re_Tropical_Storm_SENDONG_(WASHI)_Status_of_Early_Recovery_Programs_in_Region_X_issued_10FEB2014.pdf
[f]NDRRMC Update 25 December 2012. Retrieved from http://www.ndrrmc.gov.ph/attachments/article/1344/Effects_of_Typhoon_PABLO_(Bopha)_Situational_Report_No_38_as_of_25DEC2012_0600H.pdf
[g]Computed from Rappler's estimate on totally and partially damaged houses. "In Numbers: 2 Years after Yolanda". Retrieved from http://www.rappler.com/newsbreak/iq/111828-in-numbers-years-after-typhoon-yolanda
[h]NDRRMC Update 25 October 2016. Retrieved from http://ndrrmc.gov.ph/attachments/article/2946/Sitrep_No_09_re_Preparedness_Measures_and_Effects_of_Super_Typhoon_LAWIN_(HAIMA)_as_of_25OCT2016_0800H.pdf

Almost a year after, another super typhoon, Tropical Storm Nesat (locally called Pedring) hit the Philippines with a total damage of PHP 15.4 billion in infrastructure, property, and agriculture and a total number of 49,490 damaged houses: 6852 were totally damaged, and 42,638 were partially destroyed (Cf. Table 2.1). Nesat displaced 204,376 individuals due to the massive floods caused by the torrential rain, particularly in Northern and Central Luzon. A total of 13,629 families/64,431 persons (from 11,322 families/53,747 persons) were affected in 153 barangays, 25 municipalities and 10 cities in the 9 provinces of Regions I, II, III, IV-A, and V, and NCR.[3]

In a span of just 3 months, Super Typhoon Nesat was followed by Tropical Storm Washi (locally known as Sendong). Although Washi had the lowest estimated total damage, PHP 2 billion, compared with other super typhoons from 2009 to 2016 (see Table 2.1), it is nonetheless considered the deadliest super typhoon to hit the Philippines in 2011. The storm, which struck Northern Mindanao, the Visayas, and Palawan, killed at least 957 people. National Disaster Risk Reduction and Management Council (NDRRMC) Director Benito Ramos said they had lost

[3]Official Gazette (2011) "Briefer on Typhoon Pedring, September 27, 2011. Retrieved from http://www.gov.ph/2011/09/27/briefer-on-typhoon-pedring-september-27-2011/.

count of the number of people who have gone missing after the flashfloods spawned by the storm.[4] The total number of damaged houses was 51,144: 13,585 were totally damaged, and 37,559 were partially destroyed.

After a year, the deadly Super Typhoon Bopha (locally known as Pablo) struck the Southern part of the Philippines. Bopha made landfall in Mindanao on December 4, 2012, and is considered one of the worst typhoons to hit the island. It caused massive flooding and landslides and killed many people living in the provinces of Davao Oriental and Compostela Valley. Judging from its total damage and destruction, Bopha could have been the strongest super typhoon to hit the country recently until Super Typhoon Haiyan (Yolanda) came in 2016.[5] Except for Haiyan, Table 2.1 shows that Bopha surpassed other super typhoons from 2009 to 2013 in terms of total damage and number of homes destroyed. The NDRRMC reported that Bopha caused PHP 43.2 billion in total damage of agriculture, infrastructure, and property in Mindanao and damaged 216,817 homes.

Finally, Table 2.1 showed that the recent Typhoon Haiyan (Yolanda), which passed through the islands of Leyte and Samar on November 8, 2013, is thus far the strongest typhoon that hit the Philippines in the recent decade. It destroyed an estimated total of 1,012,790 homes: 550,928 houses were completely destroyed, and 589,404 additional homes were partially destroyed (Environmental Science for Social Change, 2014, p. 6). This category 5 typhoon made its first landfall over Guiuan, Eastern Samar, in the early morning of November 8, 2013, and wreaked havoc, primarily on the Visayas region, until its exit from the Philippine area of responsibility the next day. It had wind speeds exceeding 185 kph when it made landfall. Its strong winds ripped roofs off of thousands of homes and knocked down shanties, trees, power and telephone lines, and cell towers. Storm surge waves reached as high as 6–9 meters (equivalent in height to a two-storey building), claimed thousands of lives, and destroyed millions worth of properties (Commission on Audit, 2014, p. 4). The scale of the destruction of homes by Haiyan forced housing to the forefront of the national agenda and forced the government as well as private local and international disaster organizations to rethink their post-disaster recovery programs, particularly those on post-disaster housing.

2.9.2 *Post-disaster Housing in the Philippines*

It is established within the disaster-management literature that post-disaster housing is a complicated process and requires a near-impossible balancing act to satisfy all parties involved (Daly & Brassard, 2011). Large-scale disasters usually involve multiple organizations and agencies having the capacity to coordinate their efforts to manage disasters (Drabek, 1986; Drabek & McEntire, 2003; Quarantelli, 2000) with

---

[4]Malig, J. (20 Dec 2011). Sendong' world's deadliest storm for 2011. Retrieved from http://news.abs-cbn.com/nation/12/19/11/sendong-worlds-deadliest-storm-2011.

[5]Santos, R. (4 Dec 2013). "TIMELINE: Looking back at 2012s Typhoon Pablo (Bopha)." Retrieved from http://www.rappler.com/rich-media/45191-timeline-typhoon-pablo-bopha-2012.

each organization or agency having its own mandate, pre-existing social networks, and discretionary powers as to when and how it coordinates with another agency. Creating a coherent and comprehensive approach to post-disaster recovery and reconstruction after a major large-scale disaster has always been a challenge to disaster managers.

Although some international conventions and laws exist on how countries deal with post-disaster recovery, gaps in international and local legislation and law enforcement do exist, thus allowing the normative orders of those who participated in the planning and implementation of PDR programs to intervene and reshape the outcome of the intended reconstruction goals. In the international arena, there is a growing lack of legislation that governs international relief operation and donation for reconstruction. With the growing number of disasters worldwide occurring every year due to climate change, the unprecedented increase in humanitarian aid for reconstruction, particularly for post-disaster housing, is expected to increase tremendously. Because donor countries do not usually channel their donations to affected country bilaterally, it is expected that non-governmental organizations (NGOs) would receive the bulk of aid for reconstruction purposes. The volunteer organization or NGO normally disburses the funds in accordance with the organization's goals and the donor's intention. In the Philippines, some NGOs often participate in post-disaster in housing projects of the government but primarily operate within the normative mandate of their donors and organizations.

The Philippines has no comprehensive and coherent system concerning post-disaster housing and relocation for disaster victims of major natural calamities. "Although the country experienced a major disaster after another in the last decade, the strategies on post-disaster housing response continues on a disaster-event basis. In the absence of a comprehensive law on post-disaster management, "the national government and local government units have yet to standardize a clear strategy that lays down basic coordination and clear definition of roles among the national government agencies involved in post-disaster housing responses, for an efficient and integrated response…" (Environmental Science for Social Change, 2014, p. 5). "Post-disaster housing reconstruction is presently undertaken through the resettlement program led by the National Housing Authority (NHA), the core/modified shelter assistance program (C/MSAP) of the Department of Social Welfare and Development (DSWD), and the community mortgage program (CMP) of the Social Housing and Finance Corporation (SHFC)" (Miclat & Annawi, 2014). International and local as well as public and private donors also participate in a variety of ways in providing temporary and permanent shelters to disaster victims. With the absence of a comprehesive PDR legislation and one-stop post-disaster management agency that takes care of the housing needs of disaster victims, numerous deviations and negative consequences can be expected in Philippine PDR housing and resettlement.

2.10 Summary

This chapter showed that despite the advancement in theorizing and methodology in the study of disaster, the sociological perspective remains marginalized in disaster research as well as in sociology of disasters and the social sciences. It also showed that the focus of the mainstream sociology of disasters of seeing disaster events as systems, rather than an arena of conflict, has sidelined examination of the social structure and normative pluralism in disaster situations, specifically in evaluating the unintended effects of multiple legal and social normative systems surrounding the implementation of PDR projects. Thus, the normative pluralist approach of socio-legal studies and the sociology of law in evaluating the success or failure of PDR projects is apparently missing in the sociology of disasters. Yet, most disaster-prone countries in the world are found in some urbanizing countries of Southeast Asia with high-context cultures, rigid business regulation, high red-tape and corruption indices, and complex legal and social normative systems that lead to more negative unintended effects in the enforcement of PDR projects such as post-disaster housing. The Philippines—being a disaster-prone Southeast Asian country with a high level of legal and normative pluralism, red tape, corruption, rigidity in bureaucratic regulation, and personalism in contractual relations—presents a unique case in assessing the success and failure of PDR housing. The normative pluralist approach in the study of PDR process, which is rarely used in the sociology of disasters, can present as an attractive analytical tool to assess the many failures and negative unintended consequences of PDR housing projects in the Philippines.

References

Adger, W. N. (1999). Social vulnerability to climate change and extremes in coastal Vietnam. *World Development, 27*(2), 249–269.

Adger, W. N. (2000). Social and Ecological Resilience: Are They Related?. *Progress in Human Geography*, *24*(3), 347–364. doi:10.1191/030913200701540465

Alexander, D. E. (2013a). Resilience and disaster risk reduction: An etymological journey. *Natural Hazards and Earth System Science, 13*, 2707–2716.

Alexander, D. E. (2013b) Resilience and disaster risk reduction: an etymological journey. *Natural Hazards and Earth System Sciences*, *1*, 1257–1284.

Auf de Heide, E. (1989). *Disaster response: Principles and preparation and coordination*. St. Louis, MO: The CV Mosby Company.

Bankoff, G., & Hilhorst, D. (2009). The politics of risk in the Philippines: comparing state and NGO perceptions of disaster management. *Disasters, 33*, 686–704. doi:10.1111/j.1467-7717.2009.01104.x

Bildan, L. (2003). *Disaster management in southeast Asia: an overview*. Pathumthani, Thailand: Asian Disaster Preparedness Center.

Birkman, J. (ed.). (2006). Measuring Vulnerability to Natural Hazards: Towards Disaster Resilient Societies. New York: United Nations University Press. xxvi + 524 pp. ISBN 9280811355.

Birkmann, J., & Ben W. (2006). Measuring the unmeasurable. UNU-EHS SOURCE, 5. Retrieved February 23, 2014 (http://www.ihdp.unu.edu/file/get/3962.pdf)

Bodin, P. & Wiman, B. (2004). Resilience and Other Stability Concepts in Ecology: Notes on their Origin, Validity, and Usefulness. *The ESS Bulletin*, *2*(2).

Bolin, R. & Bolton, P. (1986). Race, Religion, and Ethnicity in Disaster Recovery, Monograph #42, Program on Development and Behavior. Boulder: Institute of Behavioral Science. University of Colorado.

Bonnano, G. A. (2004). Loss, trauma, and human resilience: Have we underestimated the human capacity to thrive after extremely aversive events?. *American Psychologist*, *59*(1), 20–28.

Bowen, J. R. (2005). Normative pluralism in Indonesia: Regions, religions, and ethnicities. In W. Kymlicka & B. He (Eds.), *Multiculturalism in Asia* (pp. 152–169). Oxford University Press.
Bozeman, B. (1993). A theory of government "red tape". *Journal of Public Administration Research and Theory, 3*(3), 273–304.
Brassard, C., & Raffin, A. (2011). Resilience in post-disaster societies: From crisis to development. *Asian Journal of Social Science, 39*(4), 417–424. Retrieved from http://www.jstor.org/stable/43498806
Burton, I., & Kates, R. W. (Eds.). (1993). *The environment as hazard.* New York: Guildford.
Cardona, O. D. (1993). 'Natural disasters, global change and sustainable: a strategy for reducing effects', III Meeting of the Scientific Advisory Council for the International Geophere-Biosphere Programme, Forum on Earth System Research, ICSU, Ensenada, Baja, California, Mexico.
Coleman, J. (1990). Foundations of Social Theory, Belknap Press, Cambridge, MA.
Commission on Audit (2004). Overall Report on the Sectoral Performance Audit of the Housing program of the Government. Commission on Audit Report No. 2004–02. https://coa.gov.phindex.php/sectoral-performanceaudit/category/244-housing-program-of-the-government?download=700:overall-report-ms-report-no-2004-02
Da Costa, K., & Pospieszna, P. (2015). The Relationship between human rights and disaster risk reduction revisited: Bringing the legal perspective into the discussion. *Journal of International Humanitarian Legal Studies., 6*(1), 64–86.
Crouch, E., & Wilson, R. (1982). *Risk/benefit analysis.* Cambridge: Ballinger.
Cutter, S. L., Boruff, B. J., & Shirley, W. L. (2003). Social vulnerability to environmental hazards. *Social Science Quarterly, 84,* 242–261.
Cutter, S. (1996). Societal Responses to environment hazards. *International Social Science Journal, 48*(150).
Dahlberg, R. et al. (2015). Resilience in disaster research: three versions. *Civil Engineering and Environmental Systems, 32*(1–2), 44–54.
Daly, P., & Barassard, C. (2011). Aid accountability and participatory approaches in post-disaster housing reconstruction. *Asian Journal of Social Science, 39*(2011), 508–533.
Deflem, M. (2012). Introduction: Disasters and hazards in socio-legal studies. In Deflem, M. (ed). Disasters, Hazards, and Law. Bingley, UK: Emerald Group Publishing.
Dietz, T. M., Stern, P., & Rosa, E. (1993). Risk, technology, and society. In R. E. Dunlop & W. Michelson (Eds.), *Handbook of environmental sociology.* Westport, CT: Greenwood Press.
Dirks, K. T., & Ferrin, D. L. (2001). The role of trust in organizational settings. *Organization Science, 12,* 450–467.
Douglass, M. (2013). The urban transition of environmental disasters in Asia. ARI Working Paper 2010. October 2013. Retrieved from https://ari.nus.edu.sg/Assets/Uploads/docs/wps/wps13_210.pdf
Douglas, M. & Wildavsky, A. (1982). Risk and Culture. Berkeley, CA: University of California Press.
Downing, T. E., Olsthoorn, A. A., & Richard, S. J. T. (1999). *Climate, change, and risk.* London: Routledge.
Drabek, T. E. (1986). Human Responses to Disaster: An Inventory of Sociological Findings. New York: Springer-Verlag.
Drabek, T. E. (2004). Social Dimensions of Disaster, 2nd ed. Emmitsburg, Maryland: Emergency Management Institute, Federal Emergency Management Agency.
Drabek, T. E. (2005). *Sociology, disasters and emergency management: History, contributions, and future agenda.* Retrieved from https://training.fema.gov/emiweb/downloads/drabeksociologydisastersandem.pdf
Drabek, T., & McEntire, D. (2003). Emergent phenomena and the sociology of disaster: Lessons, trends, and opportunities from the research literature. *Disaster Prevention and Management, 12* (2), 97–112. doi:10.1108/09653560310474214
Dynes, R. R. (1970). Organized Behavior in Disaster. Lexington, Mass.: D.C. Heath.
Elliot, L. (2012). ASEAN and environmental governance: Strategies of regionalism in southeast Asia: Global environment. *Politics, 12*(3), 38–57.

Environmental Science for Social Change (2014). Rapid Assessment of the Performance of Post-Disaster Housing Reconstruction Approaches: Submitted to the World Bank-Philippines Office. 30 June 2014. http://essc.org.ph/content/wp-content/uploads/2014/10/ESSC-final-report_30June-2014.pdf

Espia, J. P., & Fernandez, P. (2015). Insiders and outsiders: Local government and NGO engagement in disaster response in Guimaras, Philippines. *Disasters*, *39*(1), 51–68. doi:10.1111/disa.12086

Fallahi, A. (2007). Lessons learned from the housing reconstruction following the Bam earthquake in Iran. *The Australian Journal of Emergency Management, 22*(1), 26–35.

Farber, D. D. (2012). Introduction: Legal scholarship, the disaster cycle, and the Fukushima accident. DUKE ENVTL. *L. & POL'Y F*, 23(1). Retrieved from http://scholarship.law.duke.edu/cgi/viewcontent.cgi?article=1237&context=delpf

Fischer, H. W. (2003). The sociology of disaster: Definitions, research questions, & measurements continuation of the discussion in a post-September 11 environment. *International Journal of Mass Emergencies and Disasters, 21*(1), 91–107. Retrieved from http://www.ijmed.org/articles/82/download/

Foster, J. (1997). Successful Coping, Adaptation and Resilience in the Elderly: An Interpretation of Epidemiologic Data. *Psychiatric Quarterly*, *68*(3), 189–219.

Freudenburg, W. R. (1986). Social impact assessment. *Annual Review Sociology, 12,* 451–478.

Fritz, C.E. & Marks, E. (1954). "The NORC Studies of Human Behavior in Disasters". *Journal of Social Issues*, *10*, 26–41.

Gaillard, J. C. (2010). Vulnerability, capacity and resilience: Perspectives for climate and development policy. *Journal of International Development*, *22*(2), 218–232.

Gambetta, D. (1988). 'Can We Trust Trust?', in D. Gambetta (ed.) Trust. Making and Breaking Cooperative Relations, pp. 213–37. New York: Basil Blackwell.

Godschalk, D. R. (2003). Urban Hazard Mitigation: Creating Resilient Cities. *Natural Hazards Review*, *4*(3).

Gordon, J. (1978). Structures. Penguin Books, Harmondsworth, UK.

Guba, E. G. (Ed.). (1990). *The paradigm dialog*. Newbury Park, CA: Sage.

Guba, E. G., & Lincoln, Y. S. (1994). Competing paradigms in qualitative research. In N. K. Denzin & Y. S. Lincoln (Eds.), *Handbook of qualitative research* (pp. 105–117). Thousand Oaks, CA: Sage.

Hall, E. T. (1976). *Beyond culture*. Garden City, NY: Doubleday.

Hall, D., Hirsch, P., & Li, T. M. (2011). *Powers of exclusion: Land dilemmas in southeast Asia.* Singapore: National University Singapore Press.

Hooker, J. (2003). *Working across cultures*. Stanford University Press.

Hooks, J., & Miller, T. (2006). The continuing storm: How disaster recovery excludes those most in need. *California Western Law Review, 23*(1). Article 4.

Hummell, B. M. de L., S. L. Cutter, & C. T. Emrich, (2016). "Social vulnerability to natural hazards in Brazil". *International Journal of Disaster Risk Science*, *7*, 111–122.

Hussain, J. (2011). More than one law for all: Legal pluralism in southeast Asia. *Democracy and Society*, *7*(4), 374–389 (Legal Pluralism and Shari'a Law). http://dx.doi.org/10.1080/17419166.2011.617621

Inglehart, R., & Welzel C. (2005). *Modernization, cultural change and democracy: The human development sequence*. Cambridge, UK: Cambridge Unviersity Press.

Iversen, R. R., & Armstrong, A. (2008). *Hurricane Katrina and New Orleans: What might a sociological embeddedness perspective offer disaster research and planning?*. Retrieved from http://repository.upenn.edu/spp_papers/137

Jegillos, S. R. (2003). *Fundamentals of disaster risk management: How are southeast Asian countries addressing these?*. Makati: International Institute of Disaster Risk Management.

Klein, R. J. T., Nicholls, R. J., & Thomalla, F. (2003). Resilience to natural hazards: How useful is this concept?.*Environmental Hazards*, *5*(1–2), 35–45. doi:10.1016/j.hazards.2004.02.001

Klein, N. (2008). *The Shock Doctrine: The Rise of Disaster Capitalism.* New York: Picador.

Kreps, G. A. (2001). Sociology of disasters. In N. Smelser, & Baltes, P. (Eds-in-chief) *Encyclopedia of the social and behavioral sciences* (pp. 3718–3721). UK: Elsevier Ltd.
Kuhn, T. (1962). *The structure of scientific revolutions*. Chicago: University of Chicago Press.
Laska, S. (1989). *Review of the book sociology of disasters: Contributions of sociology to disaster research.* In R. R. Dynes, et al. (Eds.),. *Social forces* (Vol. 68, pp. 338–339). September 1, 1989.
Levine C. (2004) The concept of vulnerability in disaster research. *Journal of Traumatic Stress* [serial online], *17*(5), 395–402. Available from: Psychology and Behavioral Sciences Collection, Ipswich, MA. Accessed February 17, 2017.
Lindsey, T., & Steiner, K. (2012). Islam, law and the state in southeast Asia: Malaysia and Brunei: Volume 3. Islam, law and the state in southeast Asia: Malaysia and Brunei: Volume 3, I. B. Tauris, 2012. Available at SSRN: https://ssrn.com/abstract=2342509
Lockwood, D. (1956). Some remarks on "The Social System". *The British Journal of Sociology, 7* (2), 134–146.
Lowrance, W. (1976). *Of acceptable risk; Science and the determination of safety*. Los Altos: William Kaufman.
Lyons, M. (2009). Building back better: The large-scale impact of small-scale approaches to reconstruction. *World Development*, *37*(2), 385–398.
Malalgod, C., Amaratunga, D., & Haigh, R. (2014). Challenges in creating a disaster resilient built environment. *Procedia Economics and Finance*, *18*, 736–744. In 4th International Conference on Building Resilience, Building Resilience 2014, September 8–10, 2014, Salford Quays, United kingdom. Available online at www.sciencedirect.com
Manyena, S. B. (2006). The concept of resilience revisited. *Disasters, 30,* 434–450. doi:10.1111/j.0361-3666.2006.00331.x
Manyena, S. B., O'Brien, G., O'Keefe, P., & Rose, J. (2011). Disaster resilience: A bounce back or bounce forward ability? *Local Environment, 16*(5), 417–424.
Manyena, S. B. (2006). The concept of resilience revisited. *Disasters*, *30*(4): 433–450. Disasters Journal. Overseas Development Institute: London.
Matyas, D., & Pelling, M. (2014). Positioning resilience for 2015: The role of resistance, incremental adjustment and transformation in disaster risk management policy. *Disasters*, *39* (SI), S1–S18.
Matyas, D. & Pelling, M. (2015a). Adaptation and transformation. *Climactic Change*, *133*(1), 113–127.
Matyas, D., & Pelling, M. (2015b). Positioning resilience for 2015: The role of resistance, incremental adjustment, and transformation in disaster risk management policy. *Disasters*, *39*, s1–s18. doi:10.1111/disa.12107
McDermott, R. (2013). Compliance with normative frameworks in disaster management: a comparative study of Indonesia and Ireland. Centre for Humanitarian Action, University of College Dublin, Ireland. Retrieved November 6, 2014 from http://www.humanitarianstudiesconference.org/fileadmin/user_upload/fe_users/ronanrua/wchs_paper315.pdf
McEntire, D. (2001). Triggering agents, vulnerabilities and disaster reduction: towards a holistic paradigm. *Disaster Prevention and Management*, *10*(3), 189–196.
Mercer, J., Kelman, I., Suchet-Pearson, S., & Lloyd, K. (2009). Integrating indigenous and scientific knowledge bases for disaster risk reduction in Papua New Guinea. *Geografiska Annaler. Series B, Human Geography, 91*(2), 157–183. Retrieved from http://www.jstor.org/stable/40205074
Miclat, S. & Annawi, D. (2014). A rapid review of post-disaster housing reconstruction approaches. ESSC. http://essc.org.ph/content/a-rapid-review-of-post-disaster-housing-reconstruction-approaches/
Mitchel, J. K. (1990). Human dimensions of environmental hazards: Complexity, disparity, and the search for guidance. In A. Kirby (Ed.), *Nothing to fear: Risks and hazards in American Society* (pp. 131–175). Tucson: University of Arizona Press.

Nasreen, M. (2004). Disaster research: Exploring sociological approach to disaster in Bangladesh. *Bangladesh e-Journal of Sociology, 1*(2).

National Research Council. (1993). *Issues in risk assessment*. Washington, DC: National Academy Press.

National Research Council (2006). 9 The present and future hazards and disaster research workforce. In *Facing hazards and disasters*. Washington D.C.: The National Academies Press.

Newell, P., Pattberg, P., & Schroeder, H. (2012). Multiactor Governance and the Environment. *Annual Review of Environment and Resources, 37*, 365–387.

Norris, F. H., et al. (2007). Community resilience as a metaphor, theory, set of capacities, and strategy for disaster readiness. *American Journal of Community Psychology, 2008*(41), 127–150. doi: 10.1007/s10464-007-9156-6

Nystén-Haarala, S., Bogdanova, E., Kondakov, A., & Makarova, O. (2015). The interplay of flexibility and rigidity in Russian business contracting: The formal and informal framework in contracting. *Lapland Law Review*, 2, 110–142 (special issue: Flexibility in contracting) (Rovaniemi, Finland).

Palliyaguru, R., Amaratunga, D., & Baldry, D. (2014). Constructing a holistic approach to disaster risk reduction: The significance of focusing on vulnerability reduction. *Disasters, 38*(1), 45–61. doi:10.1111/disa.12031

Peek, L. A., & Mileti, D. S. (2002). The history and future of disaster research. In R. B. Bechtel & A. Churchman (Eds.), *Handbook of environmental psychology* (pp. 511–524). New York: Wiley.

Perry, R. W., & Mushkatel, A. H. (1986). *Minority citizens in disasters*. Athens: University of Georgia Press.

Peñalba, L. M., Elazegui, D. D., Pulhin, J. M., & Cruz, R. V. O. (2012). Social and institutional dimensions of climate change adaptation. *International Journal of Climate Change Strategies and Management, 4*(3), 308–322. doi:10.1108/17568691211248748

Prince, S. H. (1920). Catastrophe and social change, based upon a sociological study of the Halifax disaster. Ph.D. Thesis. New York: Columbia University Department of Political Science.

Quarantelli, E. L. (1966). Organization under stress. In Robert Brictson (Ed.) *Symposium on emergency operations* (pp. 3–19). Santa Monica, California: System Development Corporation.

Quarantelli, E. L. (1987). Disaster studies: An analysis of the social historical factors affecting the development of research in the area. *International Journal of Mass Emergencies and Disasters, 5*(3), 285–310.

Quarantelli, E. L. (1996) Emergent behaviors and groups in the crisis time of disasters. In Kian M. Kwan (Ed.) *Individuality and social control: Essays in honor of Tamotsu Shibutani* (pp. 47–68). Greenwich, CT: JAI Press.

Quarantelli, E. L. (1997). Ten criteria for evaluating the management of community disasters. *Disasters: The Journal of Disaster Studies, Policy and Management, 21*, 39–56.

Quarantelli, E. L. (2000). Disaster planning, emergency management, and civil protection: The historical of organized efforts to plan for and to respond to disasters. Prelimary Paper #1. University of Delaware, Disaster Research Center. http://dspace.udel.edu/bitstream/handle/19716/673/PP301.pdf?sequence=1

Quarantelli, E. L. (2005). Catastrophes are different from disasters: Understanding Katrina. At http://understandingkatrina.ssrc.org/quarantelli

Roosli, R., & O'Keefe, P. (2011). An evaluation of barriers to implementing disaster planning and the housing program in Malaysia. *Risk Management, 13*(4), 209–227. Retrieved from http://www.jstor.org/stable/41407074

Saban, L. I. (2015). Entrepreneurial brokers in disaster response network in Typhoon Haiyan in the Philippines. *Public Management Review, 17*(10), 1496–1517. doi:10.1080/14719037.2014.943271

Sartori, P., & Shahar, I. (2012). Legal pluralism in Muslim-majority colonies: Mapping the terrain. *Journal of the Economic and Social History of the Orient, 55*(4/5), 637–663. doi:10.1163/15685209-12341274

Saunders, G. (2004). Dilemmas and challenges for the shelter sector: Lessons learned from the sphere revision process. *Disasters, 28*(2), 160–175.

Schneider, S. (1992). Governmental response to disasters: The conflict between bureaucratic procedures and emergent norms. *Public Administration Review, 52,* 135–145.

See, J., & Porio, E. (2015). Assessing social vulnerability to flooding in metro Manila using principal component analysis. *Philippine Sociological Review, 63*, 53–80. Retrieved from http://www.jstor.org/stable/24717187

Sonn, C., & Fisher, A. (1998). Sense of community: Community resilient responses to oppression and change. *Journal of Community Psychology*, *26*, 457–472.

Stallings, R. A. (1978). The structural patterns of four types of organizations in disaster. In E. L. Quarantelli (Ed.) *Disasters: Theory and Research* (pp. 87–103). Beverly Hills, California: Sage.

Tamanaha, B. (2008). A holistic vision of the socio-legal terrain. *Law and Contemporary Problems, 71*(2), 89–97. Retrieved from http://www.jstor.org/stable/27592239

Tas, N., Cosgun, N., & Tas, M. (2007). A qualitative evaluation of the after earthquake permanent housings in Turkey in terms of user satisfaction—Kocaeli, Gundogdu Permanent Housing Model. *Building and Environment, 42*(9), 3418–3431.

Tierney, K. (2007). From the margins to the mainstream? Disaster research at the crossroads. *Annual Review of Sociology, 33*, 503–525. Retrieved from http://www.jstor.org/stable/29737773

Tierney, K. (2012). Disaster governance: Social, political, and economic dimensions. *Annual Review of Environment and Resources, 37,* 341–363. doi:10.1146/annurev-environ-020911-095618

Tierney, K. J. (1999). Toward a critical sociology of risk. *Sociological Forum, 14*(2), 215–242.

Tobin, G. A. (1999). 'Sustainability and community resilience: The Holy Grail of hazards planning?' *Global Environmental Change Part B: Environmental Hazards*, *1*(1), 13–25.

Torry, W. I. (1979). Anthropological studies in hazardous environments: past trends and new horizons. *Current Anthropology*, *20*(3), 517–540 (2014), The concept of resilience to disasters, In Jonas Joerin, Rajib Shaw, Ramasamy R. Krishnamurthy (Ed.) Building Resilient Urban Communities (Community, Environment and Disaster Risk Management, Vol. 15) Emerald Group Publishing Limited, pp. 35–48.

UNISDR. (2004). *Living with risk: A global review of disaster reduction initiatives*. Inter-Agency Secretariat of the United Nations International Strategy for Disaster Reduction, Geneva.

Valte, M. (2002). Chapter 6: A house for every Filipino: A pipedream in the land of promises? In *Book: Philippine Governance Report: Case studies in the management of power*. Quezon City: United Nations Development Programme, Ateneo School of Government, and Ateneo Center for Social Policy and Public Affairs. Retrieved from http://www.ombudsman.gov.ph/UNDP4/wp-content/uploads/2013/01/Chap6.pdf

Van Niekerk, D. (2012). *Transdisciplinarity: The binding paradigm for disaster risk reduction*. Retrieved from https://dspace.nwu.ac.za/bitstream/handle/10394/8572/Van_Niekerk_D.pdf?sequence=1

Weichselgartner, J. (2001). Disaster mitigation: The concept of vulnerability revisited. *Disaster Prevention and Management., 10*(2), 85–94.

Wisner, B., Blaikie, P., Cannon, T., & Davis, I. (2004). *At risk: Natural hazards, people's vulnerability, and disasters* (2nd ed.). New York: Routledge.

World Bank. (2009). *Philippines—Typhoons Ondoy and Pepeng: Post-disaster needs assessment, Main report*. http://siteresources.worldbank.org/INTPHILIPPINES/Resources/PDNAVol1MainReport.pdf

Zinn, J. O. (2009). The sociology of risk and uncertainty: A response to Judith Green's 'Is it time for the sociology of health to abandon "risk"?'. *Health, Risk & Society, 11*(6), 509–526. doi:10.1080/13698570903329490

Zinn, J. O. & Taylor-Gooby, (2006). Risk as an interdisciplinary research area. In: Taylor-Gooby, Peter und Jens O. Zinn (Ed.) *Risk in Social Science* (pp. 20–53). Oxford: Oxford University Press.

Chapter 3
Typhoon Ketsana and Normative Pluralism in the Philippine Post-disaster Recovery System

Abstract This chapter explained the post-disaster recovery (PDR) process in the Philippines under the Philippine Disaster Risk Reduction Act (PDRRMA) of 2010, the country's primary disaster-management law enacted by the Philippine legislature after the devastating Tropical Storm Ketsana (locally known as Typhoon Ondoy) that left thousands of poor people homeless in 2009. Unlike the old disaster-management law, PDRRMA does not only reorganize the country's disaster-management system but also adopt a more proactive and holistic stance to disaster management by incorporating, for the first time, some legal provisions on post-disaster recovery under the "build back better" (BBB) principle that requires adequate housing and relocation to disaster victims. However, this legal development did not assure a rule-based implementation of the PDR of typhoon victims in terms of housing and relocation. Using the sociological and normative pluralist perspectives, this chapter argued that the broad and brief provisions of PDRRMA on PDR allowed the influence of the plurality of state laws and regulations, as well as informal and cultural normative systems, to determine the outcome of the Typhoon Ketsana victims' PDR, resulting in more negative unintended effects that intensified the suffering of the disaster victims.

Keywords Typhoon Ketsana · Post-disaster recovery · Legal pluralism · Normative pluralism · Corruption · Post-disaster housing

3.1 Typhoon Ketsana and the Philippine PDR Legislation

One of the most destructive typhoons that ever hit the Philippines in 2009 is the Tropical Storm (TS) Ketsana, locally known as Typhoon Ondoy. The country's National Disaster Coordinating Council (NDCC)[1] reported that with TS Ketsana's enhanced monsoon rain pouring down more than a month's rain in 6 hours, widespread flooding occurred in almost all parts of Metro Manila and some parts of

[1]The National Disaster Coordinating Council (NDCC) was renamed the "National Disaster Risk Reduction Management Council" after the enactment of PDRRMA in 2010.

V.O. Ballano, *Law, Normative Pluralism, and Post-Disaster Recovery*,
DOI 10.1007/978-981-10-5074-9_3

Luzon, Visayas, and Mindanao, eventually affecting 4,901,234 persons in 786 barangay, killing 464 people, and leaving a damage of 11 billion pesos. TS Ketsana, which hit the Philippines on September 26, 2009, was a different type of storm. It brought an enhanced monsoon rain, which caused the most massive flooding in the country in 40 years. It injured hundreds of people and destroyed an estimated PHP 11 billion worth of infrastructure and farming. It also left thousands of poor families residing in low-lying areas and danger zones homeless and waiting for the urgent long-term post-disaster recovery (PDR) response from the government in 2009.[2]

Realizing the government's incapacity to address Ketsana's unexpected damage and reconstruction requirements, the Philippine legislature immediately amended its old disaster law and enacted a more proactive statute to reorganize the government's disaster-management team from the national to the local government level, mitigate disaster risks by introducing new response measures, and address the post-disaster recovery needs of victims by applying the universal principle of "build back better" (BBB).[3] Thus, on May 29, 2010, the Fourteenth Philippine Congress approved the Disaster Risk Reduction Management Act of 2010 or Republic Act No. 10121 (PDDRMA), thus updating the old disaster-management law or Presidential Decree (P.D.) No. 1566 of 1978, and changed its disaster-management approach from a reactive emergency and disaster response to proactive disaster risk–reduction management and post-disaster recovery in compliance with international standards.[4] Although its focus is on disaster-risk reduction, the PDRRMA nevertheless addresses, for the first time, the issue of post-disaster management for disaster victims.

---

[2]*See* "Too Much Rain Too Soon" at http://newsinfo.inquirer.net/inquirerheadlines/nation/view/20090927-227074/Too-much-rain-too-soon. The enhanced southwest monsoon brought by Typhoon Ondoy caused widespread flooding in almost all parts of Metro Manila Central and Southern Luzon and some parts of Visayas and Mindanao. A total 993,227 families/4,901,234 persons were affected in 2018 barangays, 172 municipalities, and 16 cities of 26 provinces of various regions in the Philippines: 464 dead, 529 injured, and 37 missing. The estimated cost of damage to infrastructure and agriculture amounted to PHP 11 Billion. See: National Disaster Coordinating Council (NDRRC) Report at http://www.ndrrmc.gov.ph/attachments/article/92/Narrative_Report_re_Tropical_Storm_Ondoy_%28.KETSANA%29_and_Typhoon_Pepeng_%28 PARMA%29_2009.pdf.

[3]Before the enactment of the DRRM, the Philippines concurred to the adoption of the Hyogo Framework for Action 2005–2015: Building the Resilience of Nations and Communities to Disasters in January 2005. See: http://www.drr-law.org/resources/Philippines-Desk-Survey.pdf. The slogan "Build Back Better" first emerged during the multi-national recovery effort after the Indian Ocean Tsunami (Clinton, 2006; Lyons, 2009) as the need to improve current reconstruction and recovery practices and generate safer communities emerged (Mannakkara & Wilkinson, 2014). Boano (2009), Khasalamwa (2009) and Ozcevik et al., (2009) proposed that "build back better" in the post-disaster reconstruction phase must not only restore communities to their pre-disaster states but also create safer, more sustainable and resilient communities.

[4]http://www.drr-law.org/resources/Philippines-Desk-Survey.pdf.

Fig. 3.1 Satellite image of TS Ketsana (Ondoy) making a landfall in the Philippines last September 26, 2009. *Source* http://weather.com.ph/typhoon/climatology

3.2 The PDRRMA and Post-disaster Recovery

The disaster-management cycle usually has four phases: mitigation, preparedness, response, recovery, and mitigation (Rubin, 1991). And of these four phases, recovery is the most poorly understood and implemented phase in disaster management (Berke et al., 1993) as well as the least researched phase in disaster management (Barton, 1969; Rubin, Saperstein, & Barbee, 1985; Schwab, 1998). Recovery can be defined in different ways. However, it is generally understood as the long-term activities undertaken to recover from a disaster event in an attempt to return the community to its pre-disaster norms (Joakim, 2008). The old Philippine disaster-management law (P.D. 1978) was concerned with disaster preparedness and emergency response and not with disaster-risk reduction and PDR of disaster victims. Thus, the passage of the PDRRMA was hailed by the Philippine government as a landmark legislation that transformed Philippine disaster management from a reactive to a more proactive disaster-response system. This law is said to have provided a holistic framework for safer adaptive and resilient Filipino communities toward sustainable development through the four (4) interrelated thematic areas of disaster management, namely, prevention and mitigation, preparedness, response, and rehabilitation and recovery.[5]

Adopting some of the United Nation's (UN) provisions on disaster management, the PDRRMA included, for the first time, some legal norms on the long-term PDR

[5]Philippine Statement (Delivered by Senator Loren Legarda on behalf of President Benigno Aquino III). 3rd UN World Conference on Disaster Risk Reduction Sendai, Japan, March 15, 2015. http://www.senate.gov.ph/press_release/2015/0315_legarda3.asp.

of disaster victims, albeit in a brief and broad manner.[6] It defines PDR as the "restoration and improvement, where appropriate, of facilities, livelihood and living conditions of disaster-affected communities, including efforts to decrease disaster risk factors, in accordance with the principles of 'build back better'" (BBB) (Section 3, [aa]). PDRRMA's legal provisions demand that the PDR of disaster victims must include adequate housing and relocation under the international principle of BBB. Specifically, they require the government to "[p]rovide maximum care, assistance, and services to individuals and families affected by the disaster, implement emergency rehabilitation projects to lessen the impact of the disaster and facilitate the resumption of normal social and economic activities" (Section 2, [p]).

3.3 PDRRMA's PDR Standards

PDR may be understood in different terms. In its short term, as contemplated by the PDRRMA, it implies that the government must restore the basic services and provision of temporary housing to disaster victims in evacuation centers. In its long term, it requires—according to the UN and international principles—that the government must provide adequate relocation and housing to disaster victims with efforts to return the community to conditions that existed before an event or even better, or ideally, to a condition that improves the social, economic, and natural environments (Flatt & Stys, 2013).[7] The PDR, as understood by the PDRRMA for disaster victims, covers these two dimensions and requires the following basic components: respectable temporary evacuation for short-term PDR and adequate housing and relocation for long-term PDR under the BBB principle. To the homeless and poor Typhoon Ketsana victims, this recovery implies resettlement to a safe location with adequate housing, social services, and livelihood opportunities. The right to adequate relocation and housing of disaster victims as defined by the Office of the United Nations High Commissioner for Human Rights (OHCHR) and tacitly adopted by the PDRRMA does not only include "a house with four walls and a roof" but also the occupants' right to security of tenure to their housing units, the

[6]United Nation Resolution No. 46/182, issued during the 78th plenary session on 19 December 1991, recognized the need for countries to adopt a framework for disaster reduction, mitigation, and the prevention of and preparation for disaster risks. Countries adopted this resolution during the World Conference on Disaster Reduction in Kobe, Hyogo, Japan, Countries in the Hyogo Framework for Action on Disaster Mitigation. The members of the Association of Southeast Asian Nations (ASEAN) also adopted this framework on 25 July 2005 by signing the Agreement on Disaster Management and Emergency Response (AADMER) (Calde, 2013). The Philippines, being a member of the ASEAN, passed two disaster laws to legislate this framework: the Climate Change Act (Republic Act 9279) and the Disaster Risk Reduction Management Act (Republic Act 10121).

[7]Although not understood within the context of disaster, Tercan (2001) defined relocation as removal to another location due to the provision of land or housing either voluntarily or involuntarily (as cited in Imura & Shaw, 2009, p. 9).

affordability and habitability of the homes, as well as the suitability of the housing location, i.e., the resettlement must not be situated in polluted or dangerous areas or be cut off from employment opportunities, health-care services, schools, childcare centers, and other social facilities (Cf. OCHR, 2009, pp. 3–5).[8]

3.4 Normative Pluralism and the Philippine PDR System

Normative pluralism in PDR can refer to the multiple, overlapping, and plural normative orders embracing formal and informal legal regimes, customs, and practices in the formulation and implementation of PDR projects (Sheppard et al., 2013). A normative order or normative control system can refer to a set of related commands, injunctions—i.e., "dos and don'ts"—that stem from the same source of a multitude of similar sources (Klabbers & Piiparinen, 2013). Normative pluralism includes the multiplicity of laws, rules, regulations, and informal cultural norms in a given social field. The effect of pluralism in normative orders that implement complex PDR projects is often overlooked in the current research and literature on post-disaster management research when assessing the failure or success of PDR housing projects, especially government projects, which are expected to be above board and guided by the "rule of law" as mandated by the state's administrative code.

The quantity and quality of normative standards used by implementers behind a project can also be considered crucial in determining the failure or success of a government's PDR programs. The greater the number of state laws and regulations and informal normative structure of the various agencies and organizations that interpret and implement the official PDRRMA, the greater is the probability of polycentric enforcement of the PDR provisions. This situation can allow some powerful interest groups to exploit the PDR system resulting in multiple negative unintended consequences that are detrimental to the welfare of disaster victims. The Philippines has no unified legal system that deals directly with the rehabilitation of disaster victims. The laws, rules, and regulations that implement PDRRMA's PDR programs are taken from various pre-existing non-disaster relocation and housing legal systems of the government's different national and local agencies and their social networks.

Moreover, there is also no single government agency that directly handles the PDR of disaster victims in the Philippines despite the insistence of disaster managers on the necessity of a "one-stop shop" PDR agency for large-scale disasters to lead the entire recovery process and to answer all issues related to recovery (ADB, 2015). Long-term PDRs in the Philippines require the participation of the different

[8]The PDRRMA incorporated internationally accepted principles and guidelines of disaster-risk–reduction management, see: Rule 1, Section 3 (c) of the Implementing Rules of PDRRMA issued by the National Disaster Coordinating Council (NDCC) at http://www.ndrrmc.gov.ph/attachments/article/95/Implementing_Rules_and_Regulartion_RA_10121.pdf.

public, private, and volunteer organizations and agencies, which are not necessarily of similar political persuasion or acquainted with disaster management. Each agency or organization has its own unique set of informal cultural normative standards that interact with the multiple state laws, rules, and regulations to implement the PDR project. These numerous and discordant relocation laws and normative systems, as well as the decentralized post-disaster governance in the Philippine bureaucracy, can largely lead to inadequacies or "failures" of the long-term PDR resettlement project for disaster victims.

Aside from the multiplicity of laws and government agencies implementing PDR projects, the quality of the cultural normative orders of the various participating agencies, organizations, and groups that interpret and implement the project's legal normative order can be also crucial for the success or failure of a PDR program. Socio-legal thought has emphasized that the formal legal normative order can no longer claim automatic supremacy in the social order. Formal legal principles may be altered, manipulated, elaborated, or ignored by the social actors who give them life (Edelman and Suchman, 1997). Actors and groups can pick and choose which code of conduct and normative structure to follow in a given situation. Thus, if informal normative orders that implement the project are not aligned and supportive of the official PDR goals under the PDRRMA, more negative effects can also be expected in the outcome of the project. Specifically, if some of the informal normative orders that support a PDR project are criminologic or "corrupt," then the result of the project would be significantly different from the intended goals. A relocation developer, for instance, could lower the price of their new housing units. But this price decrease can result in the use of substandard construction materials to recover the cost of bribes given to some unscrupulous public officials.

To some stakeholders, certain acts can be illegal in pursuing a project, but they can nonetheless justifiable on the basis of some social norms or shared cultural practice (Klabbers & Piiparinen, 2013).[9] In the Philippines, a low-cost housing and relocation project is a multi-billion-peso business for private developers, contractors, suppliers, and their networks and greater corruptive opportunities for some corrupt public officers. To expedite government transactions and maximize business profits, some project implementers resort to informal and corrupt practices facilitated by some shared Filipino normative values of *pakikisama (group solidarity), lagay (bribe), tongpats (price padding), palakasan (patronage)*, and other cultural instrumental values. Some resort to informal rules to short-cut the bureaucratic requirements in housing-relocation projects, whereas others engage in direct and indirect bribery (*padulas*) to expedite the procurement of permits, licenses, and clearances.

---

[9]Klabbers, J. and Piiparinen, T. (n.d.) Normative pluralism: An exploration. Available at: www.disciplinas.stoa.usp.br.

3.4.1 Pluralism in State Laws

In an attempt to understand the nature of law in society, two major groups of legal scholars have emerged in legal history: the formalists and the realists. The formalists generally look at the law as having its own internal logic and system and are therefore independent of the influences of society. The realists, particularly the social science group, however, examine the actual operation of legal rules as influenced by social settings. Using the methodologies of the emerging social sciences, they look at the law as it operates in its actual social context or "law in action" (Quevedo, 1985). With regard to state legislation, the formalists assert the integrity, coherence, and uniformity of the law. They see statutes as internally self-consistent, integrated, and logical systems that are free from the influence of other normative systems in society (Griffiths, 2002). The realists, however, challenged this view and asserted the vulnerability of the law to the influence of other normative fields in society.

A group of socio-legal scholars, called "legal pluralists," challenged legal monism and the monolithic conception of the legal system by the formalists and saw a diversity of state and non-state laws and quasi-legal orders, in particular, in terms of geographical setting.[10] Some also "challenged the inherent superiority of statutory law and argued that societies are characterized by plural sets of norms and values" (Bavinck & Jyotishi, 2014, p. 2). Thus, legal pluralism is understood as referring to "situations where more than one legal regime is operating in a social field" (Griffiths, 1986; von Benda-Beckmann & von Benda-Beckmann, 2007; Wiber & Reccia, 2010). It can be operationally defined as "the existence of bodies of laws within the same sociogeographical space that compete for the loyalty of a group of people subject to them" (Prill-Bret, 1994, p. 687). Pluralism of laws is inevitable because various laws, legal orders, and quasi-legal orders can co-exist in a given complex society. Legal pluralists identified two configurations of legal system co-existing in one state: the local legal systems and the state law.

Various types of the legal system existing in one society often breeds varying and conflicting legal cultures, contrasting attitudes, and differing responses of people to the law. The existence of various regimes of property laws in the same state, in particular, could generate the opposite effect and thus make the legal system unnecessarily complex and force citizens to obey a large number of contradicting laws and promote differing courses of action with respect to problems related to property (Maldonado, 2009). Implementing government programs that concern property rights, such as the post-disaster housing and relocation of disaster victims, in a legal pluralist environment can lead to conflicting results. It is expected that certain normative standards emanating from local legal systems which intertwine with cultural normative systems, can challenge the official PDRRMA on PDR

[10]Legal monism is the idea that there must be one and only one centralized hierarchal legal system in each state (Maldonado, 2009).

and create negative unintended effects to the recipients of government PDR projects.

The plurality of laws and juridical fields tends to overlap and compete as well as collide with other non-legal or cultural normative structures in formulating and implementing a complex government PDR project. Legal pluralism does not only exist within one legal system but also within one specific juridical field. A legal field may not be a single but rather an aggregate normative system. The legal field of the PDR system in the Philippines under PDRRMA is polycentric. To fully implement its PDR projects, the PDRRMA must interact with and rely on various affiliate state laws, rules, and regulations of different government agencies and their network of organizations in the low-cost housing and relocation sector under the Urban Development and Housing Act of 1992 (UDHA). Chapter 4 of this book provides an overview of the major state national and local laws, rules, and regulations, as well as their respective agencies, that govern the post-disaster relocation and housing of Typhoon Ketsana victims in Rodriguez, Rizal. It attempts to illustrate how this formal legal pluralism can contribute to the inadequacies of the post-disaster relocation for these victims in Chaps. 5–8 of this book.

3.4.2 *Pluralism in Informal Normative Orders*

A loosely cohesive state laws in the implementation of a complex government long-term PDR project can often lead to interference of various informal cultural normative standards of participating agencies and groups to fill the gaps in the project enforcement. Despite the strong injunction of the legal codes for public officers and collaborators to follow official rules in government and to penalize deviations from the official prescriptions and regulations, informal normative standards oftentimes compete with the official law in actual social practice in the implementation of public projects. Although "[l]aw is all over" in society (Sarat, 1990) and has the power to constitute hierarchies and relationships that are at the heart of the social fabric of society (Yngvesson, 1993, p. 120), it can also be challenged, however, by quasi-legal orders and other non-legal normative regimes. Moore (1978) argued that it is inherent in the nature of legal systems that they can never become fully coherent, gapless, and consistent wholes that can successfully regulate the entire fabric of social life. Rule-systems invariably include ambiguities, inconsistencies, gaps, and conflicts (Moore, 1978, p. 3). A legal regime can control certain types of behavior but not the aggregate of behavior in society (Moore, 1978, p. 30). Thus, pluralism in the Philippine PDR system does not only consists of multiple, conflicting, and overlapping laws, rules, and regulations but also nonlegal and informal normative orders of participating groups and individuals.[11] The plurality of informal normative orders

[11]Legal pluralists continue to debate the nature of law and what separates it from non-legal normative systems. In this book, laws, rules, and regulations are generally understood as official norms or laws that are enacted by the state either through the Philippine Congress, local councils

corresponds to the plurality of agencies and groups involved in the project and the complexity of the PDR legal normative orders. Normative pluralist scholars see legal pluralism as only one species of the various pluralities of the normative orders in society. They argued that pluralism does not only occur in state and customary laws and quasilegal orders but also in informal and cultural normative structures between the official law and empirical reality or what Moore calls "intermediaries" or "semi-autonomous social fields" (Moore, 1973). People are socialized from childhood to follow a variety of religious, cultural, social, professional, and legal normative systems. Together with the official law, these nonlegal norms compete and, at times, conflict with the individual or group's attention, decision-making, and action in daily life. The factors that determine what individuals would follow or disregard include the source and hierarchy of the competing norms, the form and severity of the sanction people expect to suffer for favoring one norm over another, and the likelihood of the application of the sanction (An-Naim, 2014, p. 791). These semi-autonomous fields are neither totally controlled by legal fields nor free from its influence. For Moore (1973), the semi-autonomous field is defined and its boundaries identified not by its organizations but by a processual characteristic and by the fact that it can generate rules and coerce or induce compliance from people in competition with the law. If effectively strong, these semi-autonomous social fields can undermine the official law. The social arena in which a number of government and private corporate groups deal with each other to provide low-cost housing and relocation for disaster victims in the Philippines, for instance, can constitute the various semi-autonomous social fields. Each of these fields can set its own informal rules on "how things are done" within and outside their own networks as well as on how to deal with or circumvent the official housing laws.

3.4.3 *Cultural Normative Orders*

Studying the cultural values of people in organizations and projects is gaining prominence in scientific and public discourse and social research. Values are believed to have a substantial influence on individual and organizational behavior in project management and thus can constitute normative orders that may compete with the state laws (Feather, 1995). Cultural values are basically classified into two types: (1) the values people place on an object or outcome (e.g., value people place on pay) and (2) the values used by people to describe a person as opposed to an object (Feather, 1995). The former are usually called "goals" or "terminal values," whereas the latter are called "instrumental values."

(Footnote 11 continued)

or by executive departments tasked by the national legislature to draft the implementing rules and regulations (IRR) of a statute.

Terminal values are self-sufficient end-states of existence that a person or group strives to achieve. However, instrumental values are modes of behavior that facilitate the attainment of terminal values (Meglino & Ravlin, 1998, p. 353). Instrumental values specify an individual's belief about how he or she "should" or "ought" to behave in actual social practice. They describe his or her internalized interpretations about socially desirable ways to fulfill his or her needs (Kluckhohn, 1951; Rokeach, 1973). They also describe how members of a group or organization should behave in order to achieve their terminal and organizational values. Instrumental values can, therefore, constitute normative orders that determine people's behavior in certain situations. Group behaviors are usually enacted within a defined context, and this context will help define one of the various possible meanings to those who are active in that context. The actual behaviors found within organizations in such a culture will probably reflect the prevailing values (Smith, Peterson, & Swartz, 2002, p. 189).

Values in organizations, such as those of the various private and public agencies in a PDR project, are derived from the larger society and culture. Cultural values, as seen among organizational participants, simply specify those values in society (Arce, 2001). Their function in an organization or unit is similar to their function in society at large (Meglino & Ravlin, 1998, p. 357). Because of their significant impact on behavior, some organizational researchers conceptualize culture in terms of observable norms and values that characterize a group or organization (O'Reilly & Chatman, 1996). They typically stress quantitative measurement schemes in examining cultural values rather than their phenomenological meaning (e.g., Rousseau, 1990).

In an organizational setting, shared cultural values can bring people together, but conflicting values can deter people from building social bonds, create an atmosphere of hostility and diversity, and result in negative organizational performance or negatively affect the goals of any organization (Caparas & Hartijasti, 2014, pp. 4–5). Thus, a shared set of cultural values in a certain social situation can form a normative system, i.e., "any system of rules and shared expectations governing a particular social situation,"[12] for a particular group or network of groups. To the Filipino anthropologist Landa Jocano, Philippine cultural values are often understood as a *pamantayan*, i.e., a guide or standard for behavior, rather than *pagpapahalaga* or valuing. Values can set internal rules and act as directive forces for individual and group behavior (Jocano, 2000). Thus, dominant Filipino cultural values, when shared and practiced by a group of people in specific social situations, can form a unique normative order, i.e., "a way of doing things" (*kalakaran*) as understood by participants.

For example, the hiring of some teachers in the Philippine Department of Education, which is said to be a fair and equal process is actually based on sponsorship, personal connection, or group affiliation rather than qualification. It, can constitute a normative system called in the Philippines as *palakasan*

[12]http://www.encyclopedia.com/doc/1O88-normativeorder.html.

(patronage) system. The *palakasan* normative system in the Philippines stresses hiring through personal connections rather than merit and objective rules and laws in low-context or individualist cultures. In *palakasan*, trust as well as resources are not easily shared with those who are not part of the family network or in-group. The awarding of housing units to disaster victims in a relocation area, for instance, may not be based on legal prescription but on *palakasan*, on the personal connection of the applicants with public officials.

A normative system can also be related to other host of Filipino cultural values—such as the value of *utang-na-loob* (debt-of-gratitude), the *Padrino* (political patronage) or *bata* (protégé) systems, and other affiliate cultural values. Unlike the legal system with its formal sanctions, informal normative systems have their own system of rewards and informal sanctions. An infraction can result, for instance, in social ostracism and deprivation of in-group's resources benefits and protection.

3.4.4 "Corrupt" Cultural Normative Orders

How a culture influences corruption has been studied lately by some social scientists. Culture is a set of beliefs and values about what is desirable versus undesirable in a community of people as well as a set of formal and informal practices to support those values (Javidan & House, 2001). Understanding the dynamics of culture is crucial in the study of corruption. One of the main analytical problems with the empirical study of corruption is its high cultural variability (Torsello, 2013). Corruption is dependent on cultural context. Thus, the study of corruption is best understood in specific social contexts rather than a cross-cultural approach that uses one analytical standard across different cultures. There is an agreement among various scholars in different fields of research that distinctive societal cultures influence a wide variety of social phenomena (Hofstede, 1983). Organizational-behaviour studies indicate that cultural values can strongly influence personal behaviours (Rokeach, 1973; House et al., 2004; Hofstede, 1983) as well as the individual's perception of ethical situations; therefore, national cultural differences are expected to influence corruption (Seleim and Bontis, 2009). And corruption sidelines legal prescriptions. M. Deflem (1995) defines corruption as the colonization of social relations in which two or more actors undertake an exchange relation by way of a successful transfer of the steering media of money or power, therebey sidestepping the legally prescribed procedure to legalize the relation (Deflem, 1995, p.1).

Culturally speaking, Filipinos are said to have a privatized view of the public. They do not think of public space as being a shared community asset that must be cared for and respected by all. This task is passed on to the government (Roces & Roces, 1985). Applying the theory of "the transitory ownership of public space," Richard Stone (1971) argued that if Filipinos occupy public space or office temporarily, they treat it as a private resource that they can use for their personal, familial, or group needs. The Filipino's concept of the community is kinship-based.

What promotes the welfare of the person's community of relatives and family members is what constitutes the "common good."

This privatized view of Filipinos of public space structures the normative orientation of Filipinos cultural values toward the in-group rather than the public good which can conflict with the merit-based and welfare orientation of the PDRRMA. A privatized view of public assets can facilitate the formation of corrupt normative systems among various organizations and groups involved in government projects, particularly in public socialized housing for calamity victims. Filipino cultural values are said to be "bipolar." They can both have positive and negative effects (Andres and Ilada-Andres, 1987). When used as a behavioral guide or "pamatayan" in overcoming the highly bureaucratic housing and relocation system in the Philippines and in distributing PDR benefits to disaster victims, some Filipino instrumental values—such as *padulas (grease money), lagay (bribe), palakasan (patronage),* and the *Padrino (political patronage) and bata (protege)* systems—can sideline the dictates of the official laws and yield negative unintended effects to the long-term PDR project.

In the arena of disaster management and relocation, intermediary normative systems can refer to informal normative orders (*kalakaran*) created by the social ties, alliances, networks, or social connections of actors and groups within one organization or agency, or between two or more organizations that form a network and express some instrumental Filipino cultural values. Formulating and implementing a complex PDR project in the Philippines involves a multiplicity of public and private organizations and their networks—both the national and local levels—as well as other pre-existing semi-autonomous normative systems relations interacting with official state laws and regulations on housing and relocation. Because bureaucratic corruption is said to be endemic in the Philippine government, some of these normative structures can constitute what Andrew Szasz (2006) calls the "criminologic regulatory structure" where large private and public corporations enter into illegal networks and alliances with state regulators resulting in lax implementation and incompetent law-enforcement system; thus, regulators intentionally allow powerful interest groups to profit in a state regulatory system (Szasz, 1986). In the Philippines, there is a popular perception that the government's low-cost housing program, as well as post-disaster housing program, are dominated by large and powerful housing developers accredited with the National Housing Authority (NHA) and supported by the prevailing policy of the government's private–public partnership (PPP) which encourages private business to invest in government projects. The entry of big business in public low-cost housing program (such as real estate developers with their preoccupation with profit and their use of a criminologic cultural normative system) can undermine the welfare orientation of PDRRMA concerning the PDR of disaster victims. These developers, with strong connections in the government's housing bureaucracy and employing some of the negative informal normative systems—such as *palakasan*, the *Padrino* system, *bata*, or *utang-na-loob*—can circumvent some housing laws and regulations to ensure greater profits for their real-estate projects.

Corruption in the low-cost housing system only reflects the country's high corruption index. In fact, Transparency International (TI), has consistently placed the Philippines on its list of most corrupt countries. Corruption committed with impunity is so ingrained in Philippine bureaucracy that 'giving commissions and percentages' to politicians and public personnel have become a 'standard operating procedure'. Orchestrated public bidding is also a common practice in awarding government contracts, such as low-cost housing projects. Bureaucratic corruption has not only tainted the top government agencies such as the Bureau of Customs or the Philippine National Police (PNP) but also the top public and private corporations in charge of the government's low-cost housing for the urban poor and PDR programs for disaster victims.

Many researchers define corruption from a behavioral perspective as the abuse of public power for private benefit (Park, 2003; Heidenheimer, 1989). To sociologists, corruption is primarily a problem of social structure and social relations rather than a problem of moral defect of individuals and groups in society. In PDR projects, corruption can be seen as a conspiracy of groups and individuals based on the cultural norm of reciprocity, understood and used in a negative sense, by unscrupulous individuals and groups and governing semi-autonomous social fields that undermine the intended effects of the official law. Corruption can be either monetary or bureaucratic. In the Philippines, corrupt social fields of some participating groups and agencies in the PDR project can generate informal rules or networks through some instrumental Filipino cultural values. "Corrupt" normative orders in Philippine private and public organizations can sidestep the legally prescribed procedures under the PDRRMA and its affiliate state laws, rules, and procedures. Social ties in a graft-laden informal normative order can include friendship, ethnicity, and kinship based on either consanguinity, affinity, or *compadrazgo*,[13] thus connecting unscrupulous individuals and public officials of different government housing agencies or private developers and housing agencies or regulatory bodies.[14] The informal normative system of "padulas" (grease money)

---

[13]It is Spanish term that refers to a system of godparenting. It is a fictive kinship system established through sponsorship in the Catholic Church's baptism, confirmation, or matrimony ceremony. Through *compadrazgo*, the child's parents and godparents—now known as *compadres* (literally "co-fathers") and *comadres* ("co-mothers")—enter into a complex relationship of rights and obligations. When in need, a family often turns to its children's godparents for assistance (http://countrystudies.us/mexico/59.htm). In the Philippines, politicians are known to sponsor mass weddings and baptism to expand their political network.

[14]Investigations and allegations of corruption and *palakasan* in Philippine housing are often connected with kinship ties. The Montalban Relocatees Association (MRA), for instance, asserted that San Jose Builders, one of the developers of the the SRHP, continues to receive multimillion housing contracts from government despite numerous complaints because the executive secretary of the president is a brother-in-law of the owner of the company. In a corruption investigation in the Philippine Senate, it turned also turned out that the President of Pag-ibig, the government provider of loans for socialized housing, is a relative of the Chief of Staff of the Vice President, who is the housing czar of the government.

"tongpats," (price padding) or "lagay" (bribe) employ some of these ties to circumvent government regulations in post-disaster housing and relocation.

In actual social practice, corporate transactions between various public and private organizations in the Philippines are often influenced by instrumental Filipino cultural values that tend to undermine official laws. "Thus the more objective, aloof and detached prescription of the law can be circumvented in favor of a more culturally desirable value of *pakikisama (*camaraderie), *kakilala* (protégé), etc. that guide people's behavior. Public bidding on government low-cost housing projects can be orchestrated or the rules on public bidding can be circumvented if certain developers or contractors are *magkakakilala* (acquaintances), *magkakamag-anak* (relatives), *magkakaibigan* (friends), or *dating magkakaklase* (former classmates) of some top officials of the NHA or government bureaucracy. The *palakasan* system is another important informal normative order based on *utang-na-loob* (debt-of-gratitude) and political patronage, which is an important political feature in Philippine politics. Political patronage has penetrated the Philippine bureaucracy and government projects as well as the post-disaster management in the country. Distribution of disaster assistance, for instance, is distributed by local officials, such as barangay officers, based on political connections and interests (Sherwood et al., 2015).

Arce (2001) argued that in Philippine organizations, personalism and familism and a host of other values—such as "smooth interpersonal relationship, *pakikisama*, *utang-na-loob*, *hiya*, and other values which are supportive of harmonious relationships such as individual's personal network of selected relatives and other allies are emphasized and the impersonal and more universal values such as merit principle and rationality of procedures demanded by official laws are not central in organizational behaviors" (Arce, 2001, pp. 187–190). Access to organizational resources for Filipinos, for instance, may not be based on merit and qualification as demanded by official laws and procedures but based on social ties and patronage (*palakasan*). The social world of Filipinos is viewed as oriented toward the welfare of a small circle of friends and relatives or the idea of *sakop* (in-group) rather than public good. This is confirmed by some studies on the social capital of Filipinos. Although studies that explore the nature of social capital in the Philippines are scarce, they do point to the narrow view of Filipinos on social networks. They also indicate the personalistic and in-group orientation of Filipino's cultural values. Ricardo G. Abad (2005), based on a national survey of Social Weather Stations (SWS), offers a more formal and quantitative analysis on the characteristics of the social capital of Filipinos. Focusing on the configurations of two key notions of social capital, namely, social networks and trust, Abad concluded that Filipinos generally build strong binding social capital with family members and depend on them for material, psychic, and symbolic needs as well as with some close friends as part of their social networks to augment what the family and kinship ties cannot or only meagerly offer (Abad, 2005: 44). These strong binding ties with their families, relatives, and friends are, however, accompanied by their relative lack of bringing social capital or ties to wider networks obtained from membership and participation in organizations (ibid). Thus, the cultural value of *sakop* (in-group), or small-group solidarity, is more important for Filipinos in resource sharing than the value of the public or common good as demanded by official PDR laws. Despite the prevalence of official laws, Filipinos remain very personalistic in their ways: To them, *mas importante ang* loyalty, commitment, *pakikisama* (loyalty, commitment, and camaraderie) are more important compared with nonpersonal traits such as efficiency or effectiveness (Franco, 1986, p. 36) as demanded by official laws.

3.5 Summary

This chapter has illustrated that the passage of PDRRMA in 2010 did not automatically guarantee a more holistic approach to PDR, especially in terms of providing adequate housing and relocation to Typhoon Ketsana victims. The implementation of PDR projects in the Philippines has to undergo the legal and normative pluralism of the country's real-estate administration which can result in more negative unintended consequences to PDRRMA's PDR housing goals. The PDRRMA's lack of legal provisions to guide actual post-disaster housing for disaster victims resulted in the reliance on the existing relocation system of UDHA and its pre-existing informal normative systems, which are governed by some negative Filipino values and corrupt practices. Given the Philippines' legal pluralism, complex real-estate regulation, and personalistic cultural values, serious deviations from the intended goals of PDRRMA are more likely to occur that could increase the suffering of disaster victims.

References

Abad, R. G. (2005). Social capital in the Philippines: Results from a National survey. *Philippine Sociological Review, 53*(1–4), 1 Jan–Dec 2005

ADB (2015) Proceedings of the regional knowledge forum on post-disaster recovery, 20–21 October 2015. Asian Development Bank. https://www.adb.org/sites/default/files/publication/176471/proceedings-forum-post-disaster.pdf

Andres, T.D., & Ilada-Andres, P.C.B. (1987). *Understanding the Filipino*. Quezon City, Philippines: New Day Publishers.

An-Naim, A. (2014). Islam and religious pluralism: It is normative, not legal? At the International Conference on Legal Pluralism sponsored by the International Association of Legal Sciences (IALS-Unesco) held at the Accademia delle Scienze di Torino in Torino, Italy (September 13, 2014). http://sbmblog.typepad.com/files/an-na_im.pdf. Accessed January 11, 2016.

Arce, W. (2001). Values, organizational culture and organizations. *Loyola Schools Review, 1*(1), 2001.

Barton, A. H. (1969). *Communities in disaster: A sociological analysis of collective stress situations*. Garden City, New York: Doubleday & Company Inc.

Benda-Beckmann, F. von & von Benda-Beckmann, K. (2007). Transnationalisation of law, globalisation and legal pluralism: A legal anthropological perspective. In: C. Antons & V. Gessner (Eds.), *Globalisation and resistance: Law reform in Asia since the crisis* (pp. 53–80). Oxford: Hart.

Berke, P. et al. (1993). Recovery after disaster: Achieving sustainable development, mitigation and equity.*Disasters, 17*(2), 93–109.

Bovinck, M. & Jyotisi, A. (2014). *Conflict, negotiation and natural resource management. A legal pluralism perspective from India.* New York: Routledge.

Caparas, M. V. & Hartijasti, Y. (2014). Filipinos and Indonesians' shame: Towards integrity and soundness in organizational performance. Asian Institute of Management. Business and Development Research. AIM working paper series 14–021. http://www.aim.edu/files/download/1207. Accessed February 12, 2016.

Edelman, L. B., & Suchman, M. C. (1997). The legal environments of organizations. *Annual Review of Sociology 23*, 479–515.

Feather, N. T. (1995). Values, valences, and choices: The influence of values on the perceived attractiveness and choice of alternatives. *Journal of Personality and Social Psychology, 68,* 1135.

Flatt, V. B. & Stys, J. J. (2013). Long term recovery in disaster response and the role of non-profits (April 22, 2012). Oñati socio-legal series, Vol. 3, No. 2. Available at SSRN: http://ssrn.com/abstract=2119879. Accessed February 1, 2015.

Franco, E. A. (1986). *Pinoy management*. Metro manila: National bookstore.

Griffiths, A. (2002). Legal pluralism. In R. Banakar, & M. Travers (Eds.), *An introduction to law and social theory* (pp. 289–310). Hart Publishing.

Griffiths, J. (1986). What is legal pluralism? *Journal of Legal Pluralism 24,* 1–56.

Heidenheimer, A. J. (1989). Perspectives on the perception of corruption. In A. J. Heidenheimer, & V. T. LeVine (Eds.) *Political corruption: A handbook* (pp. 149–163). New Brunswick, NJ: Transaction Publishers.

Hofstede, G.H. (1980) *Culture's consequences: International differences in work-related values* (revised and expanded in 2001). Thousand Oaks, CA: Sage.

House R. J. et al. (Eds.) (2004). *Culture, leadership, and organizations: The GLOBE study of 62 societies*. Thousand Oaks, CA: Sage.

Javidan, M., & House, R. J. (2001). Cultural Acumen for the global manager: Lessons from Project GLOBE. *Organizational Dynamics, 29*(4), 289–305.

Joakim, E. (2008). *Assessing the 'Hazards of Place' model of vulnerability: A case study of waterloo region*. Waterloo: Wilfrid Laurier University.

Jocano, L. (2000). *Filipino value system: a cultural definition*. Metro Manila: Punlad Research House.

Klabbers, J., & Piiparinen, T. (2013). Normative pluralism : an exploration. In Jan Klabbers & Touko Piiparinen (Eds.) *Normative pluralism and international law: exploring global governance*. Cambridge, UK: Cambridge University Press.

Kluckhohn, C. (1951). Values and value-orientations in the theory of action: an exploration in definition and classification. In: T. Parsons, & E. Shils (Eds.) *Toward a general theory of action* (p. 388–433). Cambridge, MA: Harvard University Press.

Maldonado, D. (2009). Extralegal property, monism, and pluralism. *The University of Miami Inter-American Law Review 40*(2), 213–238.

Meglino, B., & Ravlin, E. (1998). Individual values in organizations: Concepts, controversies, and research. *Journal of Management, 24*(3), 351–389.

Moore, S. F. (1973). Law and social change: The semi-autonomous social field as an appropriate subject of study 7. *Law and Society* (pp. 719–746). http://keur.eldoc.ub.rug.nl/.FILESWETE.NSCHAPTERS.CHAPTERS/2/730/730. Accessed February 10, 2013.

OHCHR. (2009). The right to adequate housing. Fact Sheet No. 21 (Rev. 1).http://www.ohchr.org/Documents/Publications/FS21_rev_1_Housing_en.pdf. Accessed 12 Feb 2015.

O'Reilly, C. A., & Chatman, J. A. (1996). Culture as social control: Corporations, cults, and commitment. *Research in Organizational Behavior, 18*, 157–200.

Park, H. (2003), Determinants of corruption: a cross-national analysis. *The Multinational Business Review, 11*, 29–48.

Quevedo, S. M. (1985). Formalist and instrumentalist legal reasoning and legal theory, 73 California Law Review, 119, 119–123. http://scholarship.law.Berkeley.edu/californialawreview/vol173/iss1/2. Accessed December 15, 2013.

Raval, D., & Subramanian, B. (2000). Effective transfer of best practices across cultures. *Competitiveness Review 10*(2), 183–191. Retrieved from http://search.proquest.com/docview/213090183?accountid=33657

Rokeach, M. (1973). *The nature of human values*. New York: Free Press.

Roces, A., & Roces, G. (1985). *Culture shock Philippines*. Singapore: Times Books International.

Rousseau, D. M. (1990). Assessing organizational culture: The case for multiple methods. In B. Schneider (Ed.), *Organizational climate and culture* (pp. 153–192). San Francisco: Jossey-Bass.

Rubin, C. (1991). Recovery from disaster. In T. E. Drabel & G. J. Hoetmer, (Eds.), *Emergency management: Principles and practice for local government*. Washington DC: International City Management Association.

Rubin, C., Saperstein, M., & Barbee, D. (1985). *Community recovery from a major natural disaster*. Monograph#41. Boulder: University of Colorado, Institute of Behavioral Science.

Sarat, A. (1990). The law is all over: power, resistance and the legal consciousness of the welfare poor. *Yale Journal of Law & the Humanities* 2(2), 6. Available at http://digitalcommons.law.yale.edu/yjlh/vol2/iss2/6

Schwab, J. (1998). *Planning for post-disaster recovery and reconstruction*. Planning Advisory Service.

Seleim, A., & Bontis, N. (2009) The relationship between culture and corruption: A cross-national study. *Journal of Intellectual Capital, 10*(1), 165–184. https://doi.org/10.1108/14691930910922978

Sheppard, A. et al. (2013) Humanitarian logistics: enhancing the engagement of local populations. *Journal of Humanitarian Logistics and Supply Chain Management, 3*(1), 22–36.

Sherwood, A., et al. (2015). *Resolving post-disaster displacement: Insights from the Philippines after Typhoon Haiyan (Yolanda)*. International Organization for Migration (IOM): The Brookings Institution.

Smith, P. B., Peterson, M. F., & Schwartz, S. H. (2002). Cultural values, sources of guidance and their relevance to managerial behavior: A 47 nation study. *Journal of Cross-Cultural Psychology, 33*, 188–208.

Stone, R. L. (1971). *'Lagay' and the policeman: A study of private, transitory ownership of public space*. Quezon City: Institute of Philippine Culture.

Szasz, A. (1986). Corporations, organized crime, and the disposal of hazardous waste: An examination of the making of a criminogenic regulatory structure. *Criminology, 24*(1), 1–27.

Torsello, D. (2013). The perception of corruption as social and institutional pressure: A comparative analysis of cultural biases. *Human Affairs 23*, 160–173. doi:10.2478/s13374-013-0117-5

Wiber, M. & Reccia, M. (2010). Legal pluralism and integrated management in saint john harbour, Canada. MAST, 9(1), 81–101.

Yngvesson, B. (1993). *Virtuous citizens, disruptive subjects: Order and complaint in a new England court*. New York: Routledge.

Chapter 4
Normative Pluralism in Philippine Housing and Relocation System

Abstract This chapter discussed the complexity of laws and informal norms as well as the many public agencies involved in doing real-estate business in the Philippines, particularly in establishing relocation and housing projects by government-accredited developers for disaster victims. It described the tedious and multiple legal and bureaucratic requirements—at both the national and local levels—in the Philippine real-estate system, which can delay and add cost to post-disaster housing and increase the burdens of housing beneficiaries. It identified the major informal normative systems and housing groups that created many negative unintended effects to relocation projects for disaster victims such as the Typhoon Ketsana victims in the Southville Rodriguez Housing Project relocation area. Using the normative pluralist perspective, it argued that in the absence of a comprehensive and unified post-disaster recovery law and one-stop public agency on post-disaster housing, the multiple laws, regulations, and informal cultural norms in the Philippine real-estate system could deviate the intended effects of the post-disaster recovery (PDR) provisions of the country's national disaster law–the Philippines Disaster Risk Reduction Act of 2010 (PDRRMA).

Keywords Legal pluralism · Normative pluralism · Relocation · Post-disaster housing · Corruption · Normative system

4.1 Complexity in Doing Real-Estate Projects in the Philippines

Doing business in the Philippines is generally time-consuming, costly, and complex due to burdensome and numerous government laws, rules, regulations, and bureaucratic requirements imposed by the government. The country is consistently ranked low in the World Bank's annual survey on the ease of doing business in the world. In its 2016 report, for example, the Philippines only ranked 103rd of 189 economies in the world despite government efforts to decrease red tape and simplify business transactions (World Bank, 2016). The Philippines is also regularly ranked

V.O. Ballano, *Law, Normative Pluralism, and Post-Disaster Recovery*,
DOI 10.1007/978-981-10-5074-9_4

low in the world concerning simplicity of business requirements in starting and maintaining a business enterprise and on the quality of bureaucratic regulation on business, particularly on the ease of securing business and construction permits, electricity, and real estate–related business requirements.

Despite the enactment of the Philippine Disaster Risk Reduction Management Act (PDRRMA) by the Philippine legislature in 2010 to address the challenges of post-disaster management during major disasters such as Typhoon Ketsana (popularly called Typhoon Ondoy), those enacting post-disaster relocation and housing projects for disaster victims still must deal with the existing complex, time-consuming, and costly government regulations. Moreover, PDR projects under the PDRRMA are still being governed by a complex pre-existing legal and non-legal normative systems of the existing low-cost housing and relocation for the urban poor. The PDRRMA inherited a low-cost housing program under an existing non-disaster law that is primarily intended for the relocation of urban poor in danger zones or whose homes were demolished by local governments—the Urban Development Act of 1992 (UDHA). The implementation of the post-disaster recovery (PDR) legal provisions of the PDRRMA requires the adoption of the pre-existing complex networks of agencies, organizations, and groups and their respective legal and informal rules and norms under the UDHA. This adoption has resulted in an intricate normative pluralism and polycentric implementation of post-disaster resettlement and housing under PDRRMA for Typhoon Ketsana victims as well as the inheritance of "corrupt" informal normative orders of some groups and government agencies exploiting the loopholes in the implementation of UDHA.

The Philippine real-estate business is known for pluraliism in official laws, rules, and regulations as well as complexity in legal and bureaucratic requirements. In fact, the Philippines' overall business environment has always been ranked low in international surveys due to complex laws, red tape, and heavy bureaucratic burdens in starting and maintaining a business. This is particularly true with regard to securing building permits and other requirements for real-estate transactions. Regarding the ease of obtaining permits, for instance, the Philippines received a low overall rank of 95th of 180 economies.[1] In obtaining a construction permit alone, the Philippines ranked among the lowest in the world, even slipping from rank 123rd in 2014 to rank 124th in 2015 out of 180 economies. A business owner or firm must undergo 24 procedures, spend a minimum of 94 days, and pay 1.2% of the total warehouse value due to various laws and legal requirements (World Bank Group, 2014). Any real-estate developer in the Philippines, whether contracted by the national or local government, that aims to build a low-cost housing project for disaster victims must pass through this highly bureaucratic real-estate system in the country, characterized by a complex system of laws, rules, and regulations that govern real-estate ownership,

[1]The numerous real-estate laws and regulations have increased the bureaucratic requirements, cost, and burdens and thus encourage real-estate developers to resort to bribery to shorten the process. This tedious system makes it vulnerable to corruption with bribe money solicited in exchange for speeding up transactions. This "hidden cost" is ultimately passed on to end-users in the form of high-priced and/or sub-standard housing products (Valte, 2013, p. 8).

real-estate transactions, and real estate/land use (CPBRD, 2011). Although the key prohibitions on the ownership and disposition of lands are found in the 1987 Philippine Constitution, the incentives and other regulatory mechanisms for the sale, purchase, lease, and ownership of different types of real estate are scattered in various statutes and regulations from national to local government levels. An inventory of the laws relevant to the public land disposition and land registration alone, for instance, revealed the existence of more than 60 laws (CPBRD, 2011).

To shorten this complex process and minimize the bureaucratic burdens, some real-estate developers and contractors of the government's resettlement and low-cost housing program resort to "short-cuts" or informal and illegal norms such as bribery and other forms of monetary and bureaucratic corruption, thus sidelining the official laws and regulations and jeopardizing the safety of disaster victims and beneficiaries of PDR housing and relocation.

4.2 Complexity in Land and Property Administration

Issues surrounding housing and land-property rights are considered the greatest regulatory barriers to shelter assistance to disaster victims. In Haiti, PDR managers faced regulatory barriers in providing shelters to the victims of the devastating earthquake of 12 January 2010 (IFRC, 2015). The Philippines, like Haiti, has a relatively extensive legal framework through its constitution, codes, and laws; basic rights are established, and rules and procedures are elaborated therein. An important aspect of the post-disaster recovery of typhoon victims in the Philippines is finding a suitable land and location for their housing project. This task must pass through the highly elaborate system of the government's land administration and management which can greatly delay the delivery of their housing units.

Conversely, the land administration in the Philippines is managed by both national and local governments. At the national level, the Department of Environment and Natural Resources (DENR) is mandated to administer, conserve, protect, and manage almost 15.88 hectares of forest lands or 52.9% of the country's total land area. Through its Land Management Bureau (LMB), it is also mandated to dispose and issue titles to another 6.0 million hectares of untitled lands. The Land Registration Authority (LRA), which falls under a different government agency—the Department of Justice—is responsible for the registration of the issued titles released by the DENR. However, both of these agencies have specific mandates in the overall administration of public and private lands; thus, they have their own set of laws, rules, and regulations regarding real-estate transactions. They also operate on totally different systems of records, information, and document-management systems and are said to be guided by outdated legal mandates such as the Public Land Act of 1936, the Land Registration Act of 1902, and the Property Registration decree of 1978 (HB826).

At the local level, Local Government Units (LGUs) also regulate the ownership and disposition of real estate in their localities by providing locality specific real-estate ordinances. They act as the key field managers of both public and private lands because they exercise direct administration, preservation, use, and allocation of land uses and resources in their jurisdictions. They also have a large stake in the effective and efficient management of land resources in their localities because a large part of local revenues they receive come from real property tax and land-related transactions (HB826).

4.3 Pluralism in Land Administration

One major area of concern in Phillipine real-estate affairs is the multiple government agencies, at both the national and local levels, that govern the land administration and real-estate transactions in the country. Land administration alone is handled by several government agencies. There are 19 public bureaus that deal with property administration. Because each bureau has its own set of laws and implementing rules and regulations aside from the new real-estate legislations that are being continually passed by the Philippine Congress, formal legal pluralism is an important characteristic of the real-estate administration in the archipelago. Real-estate transactions are managed by numerous and, at times, overlapping laws and implementing rules and regulations. This is aggravated by the fact that many of these existing legal prescriptions are outdated, conflicting, and inconsistent due to the overlapping functions and mandates of the agencies that implement them (Philippines Property Markets Scorecard, p. 7). As a result, unintended consequences—such as different standards for land valuation, overlapping land-titling processes, and the high cost of doing real-estate business in the country—can arise (CPBRD, 2011).

Legal pluralism, coupled with the absence of a unified system of laws and regulations that govern real-estate transactions, can make property administration in the Philippines complex, tedious, and costly.[2] There is no single set of codified laws and regulations controlling real-estate transactions. Moreover, there is currently no centralized registry for real property that would instantly provide the needed data on registered and unregistered real estate, real-estate transactions, and registered owners.[3] Thus, issues involving double titling, overlapping of titles, double sale of real estate, proliferation of fictitious titles, sale of unregistered real-estate condominium and subdivision projects by unscrupulous developers, encroachment, and conflict of actual land use with current zoning and other land-use regulations, among others, continue to hound the real-estate sector and its stakeholders (Philippines Property Markets Scorecard, p. 7). Moreover, land registration or the land-titling process in the Philippine is both performed judicially and administratively, a process that deviates from the common practice of other Southeast Asian countries such as Thailand, Malaysia, Indonesia, Cambodia, and Laos, in which

[2]Ibid.

[3]http://realtyprodil.com/?q=node/11.

land registration is performed through administrative processes only. The Land Registration Act and the Cadastral Act both fall under the Torrens system of titling process, and the proceedings in both acts are judicial in character (CPBRD, 2011, p. 2). Although most countries in the Asian region have managed to implement an overarching land law, often referred to as "Land Code," the Philippines still implements multiple and overlapping land decrees. As a result, the percentage share of documented lands in the Philippines is 66%, much lower compared with Thailand (80 + %) and Vietnam (90%) (CPBRD, 2011, p. 3). Table 4.1 illustrate some of the many duplications and overlaps between government agencies and real-estate laws and regulations in the Philippines.

Table 4.1 Key duplication/overlap between agencies

Duplication/overlap	Agencies involved	Source of duplication/overlap
Primary classification of public land as arable and disposable land	DENR/NAMRIA NCIP	Conflict between responsibilities in land classification as determined by EO 292 of July 1987 (instituting the Administrative Code) and Indigenous Peoples Rights Act (IPRA) 1987
Undertaking of land surveys for titling purposes	DENR (LMB); DAR Potentially NCIP	Administrative Code provides authority to both DENR and DAR to undertake land surveys; IPRA 1987 gives NCIP responsibility for the identification, delineation, and recognition of ancestral lands/domains
Approval of subdivision surveys for titling purposes (for land already titled)	LMB; LRA	Property Registration Decree (PD 1529 of June 1978, as amended) permits either LMB or LRA to approve such plans
Awarding of original private rights to arable and disposable land	DENR (Patents); DAR (CLOAs); Courts (court decrees); NCIP (CADTs)	Two titling processes (administrative, judicial), both mandated by law; legislation authorizing different forms of ownership rights in land occur by administrative process
Maintenance of independent, uncorrelated versions of cadastral maps/records	DENR (LMB); LRA	A consequence of two agencies involved in two titling processes; the practice is neither explicitly mandated nor necessitated by law

(continued)

Table 4.1 (continued)

Duplication/overlap	Agencies involved	Source of duplication/overlap
Compilation of land maps and Information	Multiple agencies	A reflection of differing agency needs for land information. But some unnecessary overlap occurs
Land valuation and related mapping for tax purposes	BIR; LGUs	Different valuation methods mandated by different property-taxation laws

Source LAMP-DENR, Institutional Arrangement Policy Study, July 2002
Please note the following abbreviation: *BIR* (Bureau of Internal Revenue), *CADT* (Certificate of Ancestral Domain Title), *CLOA* (Certificate of Land Ownership Awards); *DAR* (Department of Agrarian Reform), *DENR* (Department of Environment and Natural Resources); LMB (Land Management Bureau); *LRA* (Land Registration Authority); NCIP (National Commission on Indigenous People); and NAMRIA (National Mapping and Resources Information Authority). (CPBRD, 2011, p. 5)

The overlapping and duplication of government agencies and functions handling real-estate laws and regulations is the major obstacle in securing the necessary registrations, titles, permits, licenses, and other real-estate requirements for the establishment of resettlement and post-disaster housing to disaster victims. Complex bureaucratic requirements can result in procedural short-cuts and corruption to expedite the construction of relocation areas and housing projects for the poor and disaster victims.

The duplication of government agencies and functions can lead to multiple forms of title certificates and confusion over the various rights in land, multiple standards for land valuation, and multiple standards for land valuation (CPBRD, 2011, p. 6). Land-title certificates, patents, original Certificates of Title, Certificates of Ancestral Land Title, and other decrees can have different degrees of validity. For example, a patent is widely regarded as a lesser title than a Certificate of Title issued a judicial decree. Courts have not respected the indefeasibility of registered patents to the same degree as other titles, whereas some banks will not lend as much money to patent holders. The multiple systems and regulations used by public agencies for land valuation also create confusion for real property taxation, compensation for land acquired for public investment, and for land valuation under the Comprehensive Agrarian Reform Program (CARP). Valuations are often doubtful, incorrect, and influenced by local politics (CPBRD, 2011, p. 6). Finally, the sharing of responsibility between national and local government about real-estate taxation also create confusion in the collection of real-estate taxes and the valuation of taxes in land transfers.

4.4 Normative Pluralism for PDRRMA's PDR

Behaviors of actors and groups implementing PDR projects can be evaluated from a variety of perspectives, normative orders, or normative control systems. The legal order is distinct and is only one of the many normative orders in a society that can influence individual and group behavior (Klabbers & Piiparinen). Although the legal order is oftentimes viewed as hegemonic and sidelining other normative control systems (Smart, 1986), it is not one, monolithic, and unified system but diverse and constantly changing and interacting with other normative control systems in society. Legal pluralism can take different forms. It can mean plurality in terms of official state laws or formal legal pluralism. It can also imply a plurality of quasi-legal orders and customary laws in certain localities. In the case of implementing the post-disaster recovery (PDR) provisions of the PDRRMA, legal pluralism exists in the numerous official laws and regulations that govern housing and relocation in the government's low-cost housing program and post-disaster management. In the case of the PDR of Typhoon Ketsana victims, formal legal pluralism exists in the various national and local laws and implementing rules and regulations that assist and implement the PDR provisions of the PDRRMA.

Moreover, the pluralism surrounding Philippine PDR housing and relocation is not only in terms of formal and official state laws, rules, and regulations but also includes in terms of informal cultural normative orders among the various agencies and organizations that assist the government's PDR project. This normative pluralism that implements the PDRRMA's PDR provisions has resulted in polycentric interpretation and implementation of the law and numerous negative unintended consequences. The short and broad provisions of the PDRRMA regarding the adequate PDR housing and relocation of disaster victims and the absence of a one-stop disaster-management agency of the government result in the reliance of the PDRRMA on a wide range of state and non-state laws, as well as other non-legal and cultural normative orders of the various formal and informal or public and private organizations, to achieve its PDR objectives, particularly regarding the housing and relocation provisions of the Urban Development and Housing Act of 1991 (UDHA). The PDRRMA, therefore, competes with other diverse non–disaster-related Philippine laws under UDHA and other regulations to pursue its PDR goals.

4.5 The PDRRMA and UDHA

The multiple legal normative sources for the implementation of the PDR of disaster victims created a hybrid order that blurs the project-implementation system. Government and private agencies can alternately choose from a wide variety of legal standards—from international, national, or local—to actualize the PDRRMA's PDR projects. Although the PDRRMA explicitly aims to be holistic in its approach to disaster management, the legal provision of the text on long-term rehabilitation of disaster victims is brief. This lack of comprehensive provisions of the law on how the PDR program of disaster victims can be actualized has led to the eclectic adoption of various housing and relocation laws with diverse legal and social orientations, which leads to a phenomenon called "legal polycentricity."[4]

The complex nature of the PDR of a major disaster, such as Typhoon Ketsana, which requires the participation of various public and private agencies and their intermediary networks, at both the national and local levels, has further complicated the existing legal pluralism and polycentric direction of the PDRRMA's long-term PDR program. This complexity and polycentricity start with the complex and multiple government agencies in the housing and land-administration sectors and their real-estate laws, rules, and regulations in the government's housing sector and housing for disaster victims and government agencies.

In addition to the PDRRMA, other juridical fields and housing and relocation laws, primarily intended for the resettlement of the urban poor not affected by natural disasters, determine the post-disaster recovery of Typhoon Ketsana victims in Rodriguez, Rizal. In the absence of a comprehensive disaster system that specifies the PDRRMA's broad provisions on post-disaster housing and relocation, other existing laws that have their own distinct orientation, such as the UDHA (R. A. 7279), are used to supplement the PDRRMA's inadequate recovery provisions. Thus, the PDRRMA is only one of the many state laws that directly and indirectly deal with low-cost housing and resettlement of internally displaced people. Contrary to the formalists' claim that the state's official legal system is unified, official laws on relocation and socialized housing in the Philippines are polycentric, i.e., they come from different legal sources whose social fields do not necessarily relate to the recovery of disaster victims.

Although the PDRRMA speaks of a holistic approach to disaster management, it still relies on the non-disaster legal orientation of the UDHA as well as the overlapping and numerous Philippine laws, rules, and regulations on housing, relocation, and land administration that support it. As expressed in Section 29, UDHA's primary objective is to implement the relocation and resettlement of persons living in dangerous areas—such as *esteros*, railroad tracks, garbage dumps, riverbanks, shorelines, waterways, and in other public places as sidewalks, roads, parks, and

[4]Polycentricity is a result of legal pluralism, i.e., using legal sources from different sectors of one legal system (Hanne Petersen & Henrik Zahle eds., 1995). The principle of statutory construction, called *Enterpretare et Concordare Leges Legibus Est Optimus Interpretandi Modus*, states that every statute must be construed and harmonized with other statutes so as to form a uniform system of jurisprudence. Thus, policy implementers construed and harmonized the post-disaster provisions of the PDRRMA must be construed and harmonized with other statutes on relocation and housing of the urban poor under UDHA of 1992 despite the fact they have different orientations: The former is primarily for disaster victims, whereas the latter is for the forced relocation for victims of demolition or urban poor residing in danger zones. Furthermore, the doctrine of necessary implication states that what is implied in a statute is as much a part thereof as that which is expressed. Every statute is understood by implication to contain all such provision as may be necessary to effectuate to its object and purpose. See: http://aginglawyer.blogspot.com/2012/10/statutory-construction-made-easy-by.html. Socio-legal. Scholars, however, rejected this assumption and believe that use and hamonization of various laws of various sectors within the national legal system can only result in polycentricity and negative unintended consequences.

playgrounds" —not victims of natural and man-made disasters.[5] In this sense, the UDHA is primarily intended not for PDR but for the relocation of urban poor and informal settlers who live in danger zones and public areas or those who are forced by demolition of the local governments to leave their makeshift homes in public areas and resettle in government relocation areas.

The UDHA specifies the criteria of selecting housing beneficiaries (Section 16) and the provision of basic services (Section 21) in the relocation site despite its being a non-disaster legislation. UDHA's legal provisions are cascaded and adopted by the Local Government Units (LGUs) in their socialized housing and relocation programs in the light of PDRRMA using the LGU legal guide for relocation through the city or municipal housing and relocation office. In the Municipality of Rodriguez, for instance, the Housing and People's Development Office (HPDO), which has its own legal and social fields in interpreting and implementing the PDRRMA, used the UDHA and the LGU Guide in evaluating applications of local disaster victims applying for socialized housing at the NHA's relocation site as well as evaluating whether its relocation system is in accordance with the state law. Aside from UDHA, other national laws have established government corporations that build, manage, and provide housing loans to beneficiaries, secure and insure the mortgages of borrowers, and oversee the entire socialized housing program of the government.

Finally, some international laws on adequate housing and relocation of disaster victims are also currently used to aid the PDRRMA and UDHA in realizing its post-disaster recovery provisions. The Philippines is a signatory of international treaties and conventions on humanitarian assistance and disaster management under the United Nations. Section 2(b) of the PDRRMA explicitly states that it aims to "[a]dhere to and adopt the universal norms, principles and standards of humanitarian assistance and the global effort on risk reduction as concrete expression of the country's commitment to overcome human sufferings due to recurring disasters." By adhering to international standards, the PDRRMA recognizes the global standards on adequate housing and relocation because the Philippines is a State Party of the United Nation's Committee on Economic, Social, and Cultural Rights (ICESCR) and a signatory of international disaster laws (e.g., Hyogo Framework) on the relocation and recovery of displaced people within borders under the United Nations' "Guiding Principles on Internal Displacement". This set of international laws also specifies the PDRRMA's general provisions for post-disaster recovery and thus adds to the existing legal pluralism of post-disaster management in the Philippines.

Owing to the lack of sufficient legal provisions for the post-disaster recovery process of disaster victims, the Philippine Disaster Risk Reduction Management Act of 2010 (PDRRMA) adopted and inherited the pre-existing legal pluralism and

---

[5]Republic Act No. 7279, "An act to provide for a comprehensive and continuing urban development and housing program, establish the mechanism for its implementation, and for other purposes" is available at http://www.chanobles.com/republicactno7279.htm#UOPPRazfgwo.

diversity of formal and informal normative orders of the existing resettlement system under the Urban Development and Housing Act of 1992 (UDHA). The legal and non-legal normative plurality in the Philippine real-estate business complicates the relocation project under PDRRMA, particularly regarding the legal and bureaucratic requirements in government relocation and housing projects and the informal cultural norms of various PDR participating groups. It also created negative unintended consequences for the post-disaster recovery (PDR) of Typhoon Ketsana victims in Southville Rodriguez Housing Project (SRHP) in the Province of Rizal, Philippines. What follows is an overview of the complexity of the basic laws, procedures, and regulations, at both the national and local levels, of the government relocation and housing process. It also identifies the dominant informal normative orders of the hegemonic groups and actors in the PDR project and how they create some negative unintended effects to the PDR of typhoon victims.

4.6 Multiple Housing Laws and Government Agencies

The PDRRMA was enacted explicitly to decrease disaster risks, reorganize the disaster organization of the government, and provide post-disaster assistance to disaster victims. Thus, to realize its holistic disaster approach and PDR objectives, more specifically regarding the relocation and housing of disaster victims, the PDRRMA must rely on the complex pre-existing housing structures and agencies of the national and local governments.

Under the present set-up, the Philippine government's socialized housing program is managed by four key government corporations in accordance with Executive Order (E.O.) No. 90, series of 1986: National Housing Authority (NHA), National Home Mortgage Finance Corporation (NHMFC) or popularly known as *Pag-ibig*, the Housing and Land Use Regulatory Board (HLURB), and the Home Guarantee Corporation (HGC). All of these corporations are governed by major government legislations, and they create their own corporate laws, rules, and circulars to perform their mandate. Because of this, one can easily see the legal pluralism in state laws and rules at the national level with regard to low-cost housing for the urban poor and disaster victims. This pluralism excludes the multiple semi-autonomous informal rules and cultural standards of private businesses performing the government's housing projects.

The most important government housing corporation on which the PDRRMA must rely is the National Housing Authority (NHA). This agency, which was established by virtue of Presidential Decree (P.D.) 757 on July 31, 1975, is the umbrella agency of the government's low-cost housing programs. Its charter law later was amended by Executive Order (E.O.) Number 10 on March 26 1986, and by E.O. No. 20 on May 28, 2001, to strengthen the NHA's housing mandate and placed it under the supervision of Housing and Urban Development Coordinating

Council (HUDCC), which is the overall coordinating body for all of the government's housing agencies.[6] Under its charter, the NHA is the sole government agency tasked to engage in direct shelter production to the lowest-earning 30% of urban income–earners through slum upgrading, squatter relocation, and development of sites and services and construction of core housing units (COA Report, 2004). The NHA, as a national agency, has the rule-making power to pursue its mandate. In this sense, the PDRRMA is dependent on the NHA directives, rules, and regulations for its PDR, especially in terms of the provision on housing, resettlement, health care, educational services, and livelihood needs of disaster victims. Although the UDHA expressly states that the local government unit (LGU) has the main responsibility for providing livelihood, coordination with the NHA is necessary to achieve this (UDHA, Article VI, Urban renewal; and resettlement, Section 29, Resettlement). The LGU cannot initiate livelihood programs without the approval and support of the NHA, which establishes the socioeconomic infrastructures in relocation area and manages the housing project while the houses are still amortized by beneficiaries.

Aside from the NHA, the PDRRMA is also dependent on other public corporations established by another bodies of law in order that disaster victims in a PDR project can receive low-cost housing and secure government loans. One of these is the National Housing and Mortgage Fund (NHMF), popularly known as *Pag-ibig*. The financing of post-disaster housing of calamity victims is performed by this agency under the supervision of HUDCC. The Presidential Decree 1216 of 1977 mandated that all low-cost housing programs of the government are to be financed by the NHMF or *Pag-ibig* Fund through its implementing arm—the Social Housing Finance Corporation (SHFC), also known as the Community Mortgage Finance (CMF).[7] The NHMF, like the NHA, has its own charter that enables it to issue directives and rules in pursuance of their mandate, which can ultimately affect the capacity of the urban poor to secure loans from CMF for land acquisition and to pay for their housing units in the relocation.

To enable beneficiaries to receive socialized housing in the relocation area—in this case, the Typhoon Ketsana victims in Rodriguez, Rizal—the CMP provides the necessary loans and mortgage programs to the NHA through *Pag-ibig*. The NHA, in turn, acts as the collection agency of *Pag-ibig* for the amortization of the relocatees' housing loans. The beneficiaries pay their monthly amortization directly to NHA local offices; the Rodriguez Typhoon Ketsana victims pay directly to the NHA office located in Kasiglahan Village 1, Rodriguez, Rizal.

However, before relocation and housing units are ready for the disaster victims, other agencies, with their own set of laws and rules, contribute to the legal pluralism of the PDR system and can delay any new post-disaster housing project for disaster

[6]http://nha.gov.ph/about_us/brief_history.html.

[7]The Community Mortgage Program (CMP) is a financing scheme established in 1988 and managed by the Social Housing Finance Corporation (SHFC), a wholly owned subsidiary of the NHMF, that provides long-term mortgage loans to organized informal settler families (ISFs) for land purchase and housing development (Ballesteros, 2015).

victims. Aside from the NHA, the Housing and Land Use Regulatory Board (HULRB) must perform its function of providing technical assistance to the private developers and government agencies in identifying suitable land for relocation sites. Under Executive Order No. 648 c. 1981, the HULRB is the planning, regulatory, and quasi-judicial instrumentality of government for land-use development with a mission to promulgate and enforce policies on land use, housing, and homeowners associations. It is the lead agency that provides technical assistance to local government units in the preparation of comprehensive land-use plans; the regulation of housing, land development, and homeowners association; and adjudications of disputes related thereto.[8] In relation to relocation, the HLURB ensures rational land use according to the rules of the UDHA. It deals directly with private developers and builders of relocation sites accredited by the NHA. It provides them with licenses, clearances, and registrations for housing and subdivision projects. Thus, it determines the suitability and accessibility of the socialized housing project before it issues the necessary permits, registrations, and clearances to developers. The accredited NHA real-estate developers and contractors must pass through this agency before they build their relocation sites and construct housing units.

Finally, the Home Guarantee Corporation Act of 2000 (Republic Act No. 8763) created another national housing agency that handles the guarantees and insurance of all government housing mortgages including those of the NHA—the Home Guarantee Corporation (HGC). The HGC encourages banks and financial institutions to lend money to individual home buyers and housing developers and provides assurance to these lenders and housing investors by issuing loan and securitization guarantees. However, the NHMFC or *Pag-ibig* remains in charge of directly lending money to housing developers and individual/group borrowers, although the insurance of these loans comes from the HGC.[9]

To ensure coordination of all these major government agencies and accomplishment of the goals of the housing program of the government, the former President Corazon C. Aquino, by virtue of Executive Order No. 90, further created the Housing and Urban Development Coordinating Council (HUDCC), which was established in December 1986 as the umbrella agency or coordinating body of all government's housing projects headed by a national housing czar appointed by the President who sits as the *ex-officio* Chairman of the Board of all these major housing corporations.[10] This government agency, which is directly under the Office of the President, serves as the overseer, overall coordinator, initiator, facilitator, and evaluator of all government policies, plans, and programs for the housing sector. To perform its functions, it also issues rules and circulars to guide these four government corporations involved in housing. On 25 May 1989

[8]http://hlurb.gov.ph/about-us/.

[9]http://www.hgc.gov.ph/faqs.html.

[10]Commission on Audit (COA), Housing Program of the Government, Housing and Urban Development Council, Overall Report (Management Services, Report No.2004–02, Sectoral Performance Audit). Available at: http://www.coa.gov.ph/GWSPA/2004/HP04.asp.

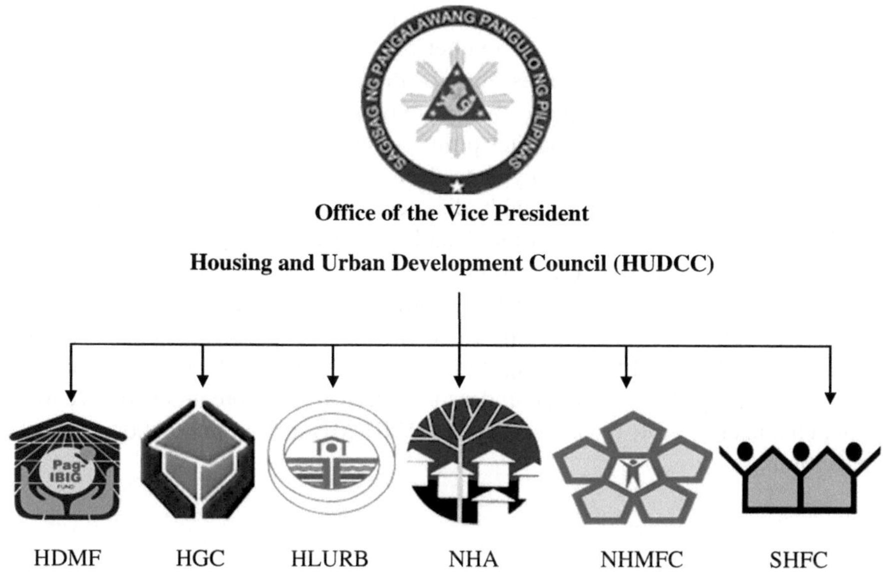

Fig. 4.1 Organizational chart of the national housing agencies under the Housing and Urban Development Council (HUDCC). *Source* www.hucc.gov.ph

and May 28, 200, Executive Order No. 357 and Executive Order No. 20, respectively, were issued to strengthen the HUDCC into a department-level organization by conferring it the power to exercise overall administrative supervision over the key housing agencies. In 1992, the Urban Development and Housing Act (RA 7279) mandated the HUDCC to direct the formulation of a National Urban Development and Housing Framework in coordination with the Local Government Units and other public- and private-sector agencies. From 1986 to the present, several executive and legislative issuances, including the three laws mentioned previously, have authorized the HUDCC with specific functions and/or required it to undertake certain tasks related to housing and urban development (Fig. 4.1).[11]

4.7 Philippine Laws Governing Relocation and Housing

Building a new resettlement site can be a very complex and slow process due to the existing laws, rules, and procedures. The current relocation and housing laws and procedures in the Philippines are clearly not intended for emergency situations such as disaster recovery. The resettlement under UDHA is apparently intended for non-emergency cases. Relocation of the urban poor by the local government due to

[11]http://www.hudcc.gov.ph/content/history.

slum upgrading or demolition is a gradual process that requires a longer time. Serving several notices and negotiation between the local government and affected communities, for instance, requires time for the government to plan the relocation and housing program for those being relocated. Authorities know that the Philippines is one of the most disaster-prone countries in the world and that a strong typhoon can hit the country at any time. Yet legal and social infrastructures have not been put in place to deal with disasters. Various supporting laws to the PDRRMA are complex, unharmonized and, at times, overlapping and in conflict with each other that cause delays and numerous negative effects to the rehabilitation of disaster victims. A review of the many national and local laws and procedures on government relocation and housing alone indicates that resettlement of disaster victims under the UDHA is a long and tedious process if official laws are to be implemented fully. The relocation under the UDHA is therefore unresponsive to the urgent character of PDR envisioned by the PDRRMA and by international housing conventions.

4.7.1 National Level

The first set of laws at the national level—with which the NHA, developers, and local government must follow to ensure the safety of disaster victims and the adequacy and suitability of the housing and relocation provision of the PDR program—deals with the building, construction, and environmental requirements and standards in setting up a new resettlement.

To ensure public safety in construction, the National Building Code (NBC) of the Philippines (P.D. 1096) is one of the most important laws that subdivision developers and contractors, including NHA-accredited developers, must comply with before commencing a relocation and housing project. The provisions of this code must be applied to the design, location, setting, construction, alteration, repair, conversion, use, occupancy, maintenance, moving, demolition of, and addition to public and private buildings and structures except for traditional indigenous family dwellings (Section 103, [a]). In particular, Section 104 of the code requires that "[a] ll buildings or structures as well as accessory facilities thereto shall conform in all respects to the principles of the safe construction and must be suited to the purpose for which they are designed. Thus, no building or construction permit will be issued by the local government if this code is not followed strictly by construction companies or developers. No resettlement site can be developed, and housing units can be constructed by the NHA and private developers without complying the basic construction standards of the NBC. The LGU where the construction of the relocation and housing are located is responsible for issuing this permit to developers."

Complying with the laws on land registration and titling of subdivision lots and housing units in the local registry of deeds is another bureaucratic burden developers must satisfy in order to divide the subdivision lots and sell them to the NHA and beneficiaries with Torrens titles annotating that the housing units were constructed on these lots.

With regard to architectural and technical specifications for building an adequate relocation site and housing project for the urban poor, the following laws are also applied in addition to the NBC and land registration and titling laws: (1) the Urban Housing and Development Act of 1992 (UDHA), (2) Presidential Decree No. 957 "Regulating the Sale of Subdivision Lots and Condominiums" (1976), and (3) its implementing rules and regulations of 2009. These laws describe the basic technical and architectural requirements for subdivision houses and structures and require developers to submit subdivision plans that comply with these requirements based on provisions of the NBC and other construction regulations to the NHA for approval. Developers must also secure registration certificates and licenses to sell (Sections 4 and 5) authorizing them to sell subdivision lots or housing units to the beneficiaries.

After the subdivision plans are approved by the NHA, the real-estate developers then submit the plans to the Director of Land for final approval in accordance with Section 44 of another law, the Land Registration Act (Act No. 496, as amended by R.A. No. 440). Batas Pambansa (B.P.) 220 and its Revised Implementing Rules and Regulations of 2009 and P.D. No. 344 (or the Accessibility Law) requiring the minimum architectural or structural requirements for enhancing the mobility of disabled persons are another set of laws that provide the different levels of standards and technical requirements for economic and socialized housing projects in urban and rural areas. They determine the technical requirements for the suitability, accessibility, and habitability of the location and housing units in socialized housing projects.

Requirements in setting up a relocation and housing for disaster victims do not usually end up following the technical and architectural standards. To ensure that the location and housing project are environmentally safe, developers must deal with new set of laws on the environment and secure permits and clearances. Presidential Decree No. 1586 and Presidential Proclamation No. 2146 and its implementing rules and regulations require an Environmental Certificate Clearance (ECC) for relocation projects.[12] Moreover, the Department of Environment and Natural Resources (DENR) Administrative Order No. 2008–20 of March 14, 2000, requires developers to comply with the guidelines on engineering geographical and geohazard assessment as an additional requirement for ECC certificates. The

[12]"[R]efers to the document issued by the Secretary or the regional Executive Director certifying that based on the representations of the proponent and as reviewed and validated by the Environmental Impact Assessment Review Committee (EIARC), the proposed project or undertaking will not cause a significant environmental impact; that the proponent has complied with all the requirements of the Environmental Impact System; and that the proponent is committed to implement its approved Environmental Management Plan in the Environmental Impact Statement or mitigation measures in the Initial Environmental Examination" (DENR ADMINISTRATIVE ORDER NO. 2000–28, Section 2,b). Available http://www.mgb.gov.ph/Files/Policies/DAO%202000-28.pdf.

rationale of this administrative order is to ensure that subdivision projects are not built in areas with geographic, geologic, tectonic, and other natural hazards such as heavy flooding, strong earthquakes, or massive landslides, which can cause great loss of life and destruction to property (Section 1). However, before developers can comply with this geohazard assessment, a certification of PHIVOLCS on active faults is needed first to identify seismic hazards. Furthermore, a conversion certification from the Department of Agrarian Reform (DAR) regional office is also needed to ensure that the land for the relocation and housing is not covered by the Comprehensive Agrarian Reform Law (R.A. 6657).

The NHA is also required to coordinate with other government housing corporations and follow with the directives of HUDCC to perform its function of providing adequate relocation and housing to disaster victims. This coordination implies a multiplicity of corporate legal orders because each of these government housing organizations follow their own legal charters and create their own internal rules and circulars to perform their respective functions. A perusal of all of these laws and legal requirements indicate that despite the legal plurality, adequate legal prescriptions that guide authorities in determining the suitability and accessibility of the relocation area, as well as the adequacy of the housing project, are confusing. Most of these laws and rules on resettlement and housing are compiled into one document at the local government level as a guide for local government units (LGUs)—the LGU's Guidebook for Local Housing Project/Program issued by the Housing and Urban Development Coordinating Council (HUDCC). Unfortunately, none has been published at the national level as an overall guide for private developers and interested parties who may wish to monitor and evaluate the adequacy of relocation and housing in NHA-owned resettlement areas. Most of the large-scale, low-cost housing and relocation project, such as the Southville Rodriguez Housing Project (SRHP) commence their projects with national housing agencies supervised by HUDCC before they are implemented in the LGUs.

The process of providing adequate housing and relocation for the PDR of disaster victims is not immune from this legal pluralism and bureaucratic burdens. Developers and housing organizations must deal with this multiplicity of rules, agencies, and bureaucratic requirements. They must also address coordination problems involving various government housing corporations and agencies and their intermediary networks before they can complete a resettlement project. This situation creates delays and thus not conducive to PDR projects for disaster victims who need an immediate response. For instance, it takes an average of one year for developers to obtain approval of permits and licenses for a relocation and housing project and another year before beneficiaries' loans can be released or taken out by the NHMFC so that the housing units can be distributed and occupied by beneficiaries (Ballesteros, 2002, p. 26). The actual occupation of the housing units by beneficiaries can even be delayed depending on the political clout of the developer with the housing bureaucracy to expedite the building process as well as the beneficiaries' connections with political patrons.

The government's direct intervention on the country's land use is primarily implemented through a land development–control system, which requires that changes in land use and the building of structures on land must pass through a system of permits and licensing (Ballesteros, 2000). The areas of government intervention include land use, subdivision development, building regulation, and environment control. To achieve this, the national and local governments require various and numerous forms of permits and licenses to all residential subdivision developers including the NHA-accredited low-cost housing developers for urban poor and disaster victims.

The five major permits that had to be secured under this process include the following:

Type	Description
Development permit	This refers to the permission granted with regard to compliance with subdivision standards and regulations based on the criteria provided under PD957 for subdivision and condominium developments or BP220 for socialized and economic housing developments
Building permit	This permit refers compliance with the National Building Code of the country. It gives the applicant the "go signal" for construction activities to start
Environmental permit	This has two forms: 1. Environment Certificate (EC): This is a certification that the project is not hazardous or has no adverse effect on the environment. It is required for all types of land development whether high-rise or subdivision developments 2. Environment Compliance Certificate (ECC): This is a requirement for projects undertaken in zones that do not conform to the approved zoning ordinances of LGUs or in areas considered "environmentally critical" by the Department of Environment and Natural Resources (DENR). The issuance of an ECC is based on systematic studies assessing the effects of a project on human health, biophysical/ecological, geophysical and socioeconomic aspects of the community
Conversion clearance	This permit is obtained when agricultural lands are to be used for urban activities. The clearance may be obtained either as an exemption or conversion certificate (AO 6 series of 1994). An exemption clearance is issued for agricultural lands classified or zoned as non-agricultural based on Town Plans approved by the HLURB before 15 June 1988 (i.e., before the CARP Law)
License or permit to sell	This permit provides the owner of the property the authority to sell any subdivision lot or condominium unit that result from a land development process. It is acquired after the development, and building permits have been issued. A license to sell also requires certifications from the utility companies (i.e. MERALCO, LWUA), a performance bond and publication. The performance bond guarantees the full development of the project

These numerous bureaucratic and legal requirements in the housing and land development process can significantly increase the costs and delays in constructing resettlement projects for PDRs. Due to the various permits, clearances, and licenses

Table 4.2 Residential subdivision development: Permits and licensing

Criteria	Year	
	1990	1998
No. of agencies	9	31
Approximate no. of approving personnel	27	155
Total no. of requirements	45	160
Time frame (months)[a]	4.5	18
No. of permits/clearances	16	14
No. of repeating requirements	6	16

[a]Time frame under normal circumstances. There are cases when the time frame is longer due to certain circumstances such as the following: (1) overlapping approved subdivision plans (LMB); (2) disagreements with farmers, tenants, or illegal claimants; (3) disagreement among LGU personnel or between the LGU and national government agencies; and (4) social pressure (Report of the Office of Special Assistant to the Executive Vice President, Filinvest, 1998 as cited in Ballesteros, 2000).

that must be obtained by developers from various government agencies, the completion of a new resettlement can take 2–3 years (Valte, 2013, p. 8) (Table 4.2).

With this complexity of requirements and delays, the Philippine government was caught off guard by the need to construct new relocations and housing units when Typhoon Ketsana (Ondoy) struck the country in September 2009. Many homeless typhoon victims waited for months in temporary shelters or evacuation tents before gaining access to the relocation sites. To expedite the construction of new houses for the numerous homeless Ketsana victims in 2010, developers had to start from scratch. To avoid long delays and red tape, they merely extended existing relocation plans, permits, and clearances of the Erap City relocation-site master plan instead of applying for new business permits, clearances, and registrations for the relocation sites of Typhoon Ketsana victims.

4.7.2 Local Level

Legal safeguards also exist at the local government level to guarantee that the relocation of and housing for the urban poor and disaster victims are both safe and adequate. The identification of sites suitable for resettlement and socialized housing is not completely determined by national agencies such as the NHA. The local government units (LGUs) also play an important role in the identification and selection of suitable sites for socialized housing. The UDHA has mandated that city and municipal governments—in coordination with the National Housing Authority (NHA), Housing and Land Use Regulatory Board (HLUB), the National Mapping Resource Information Authority (NMRIA) and the Bureau of Land (BL) —within one year after the effectivity of this law must provide updates every 3 years thereafter to conduct an inventory of all lands and improvements within their territory in order to identify suitable sites for socialized housing. After this inventory, the LGU shall identify lands for socialized housing and resettlement areas for the underprivileged and urban poor, and, under the PDRRMA, for the poor and homeless disaster victims, "taking into consideration the degree of availability of basic services and facilities, their accessibility and proximity to job sites and other economic opportunities, and the actual number of registered beneficiaries" (UDHA, Section 8).

To ensure that the UDHA provisions on site suitability for socialized housing are followed, the Housing and Urban Development Coordinating Council (HUDCC) recommended the use of the LGU's Guidebook for Local Housing Project/Program. This guidebook provides a step-by-step information on setting up a resettlement to assist LGUs in the formulation and implementation of their respective local housing programs. This document details the basic steps, procedures, and requirements in the formulation of a shelter plan including various housing schemes for the LGUs low-income constituents and implementation of local housing programs and projects on the basis of their specific mandates under the law.[13] It also contains provisions on basic services and facilities as well as livelihood program in the relocation sites. The UDHA expected the LGUs to be the main implementer of the government's relocation and socialized housing program for the urban poor after the decentralization trend in the Philippine government when it was enacted in 1992. The Local Government Code (R.A. 7160), which was enacted in 1991 to provide greater power and autonomy to local governments in provinces, cities, municipalities and barangays, mandated that LGUs must provide shelters for the homeless in their own jurisdictions. However, this legal provision has not yet been fully realized because most LGUs lack the capacity to finance and manage the relocation and housing for the homeless through their own locally-sourced funds.[14] In its latest assessment, the Bureau of Local Government Finance (BLGF), a government unit under the Department of Finance (DOF), reported that many LGUs throughout the country continued to rely on the national government's subsidy or the so-called Internal Revenue Allotment (IRA) rather than on generating local revenues for their own budgetary needs. In 2012, for instance, it reported that LGUs throughout the Philippines collected only PHP 84.57 billion from locally sourced income as opposed to the amount the IRA released to LGUs for the same year, which amounted to PHP 273.31 billion.[15]

Ironically, LGUs are in a better position to understand the socio-cultural aspects of the relocation and housing needs of their constituents than the NHA. Therefore, it is more desirable that LGUs determine the PDR and resettlement project of their own disaster victims and not the national housing corporations, such as the NHA which do not have direct contact with the homeless victims. However, the lack of

---

[13]http://www.creba.ph/images/Housing%20Library/HUDCC%20LGU%20Guidebook%20for%20Local%20/Housing.pdf.

[14]http://www.eastasiaforum.org/2010/03/27/reforming-housing-for-the-poor-in-the-philippines/.

[15]http://www.businessmirror.com.ph/3-year-review-finds-most-lgus-unable-to-raise-own-funds/.

locally sourced funds, reliance on the IRA, and the lack of capability to manage resettlements have forced LGUs to depend instead on the national government's housing program for the resettlement and housing needs of their constituents through the NHA to rehabilitate their own local disaster victims. Thus, in the case of Typhoon Ketsana victims, mayors of LGUs in Metro Manila and nearby provinces sent their own disaster victims to NHA-built relocation sites such as Southville Rodriguez. This results in off-city or near-city relocation, which detaches the disaster victims from their jobs and livelihood and encourages them to sell or lease their housing units to return to their old location and resume their livelihood activities.[16]

Sanggunian (Council) ordinances on disaster (city or municipal councils or legislatures) can also affect PDR projects. On the recommendation of the Local Disaster Risk Reduction and Management Councils (LDRRMCs) in preparing for, responding to, and facilitating recovery from the effects of any disaster, the local council or sanggunian can decrease property taxes and interest rates after a calamity. It can also declare a relocation area as unfit for resettlement. In Rodriguez, Rizal, for instance, a portion of the Kasiglahan Relocation area was declared in an ordinance by the local council as unfit for habitation after it was hit by a massive flooding.

4.8 Stages of Resettlement Projects

The intricacy of the legal and bureaucratic requirements in obtaining construction permits, clearances, registrations, and other real-estate requirements, constitutes only a small portion of the much more complex process of finishing a resettlement project for the rehabilitation and housing of disaster victims under the UDHA. Completing the major stages of the resettlement project is even more legally tedious, time-consuming, and costly. Under existing UDHA and NHA laws and procedures, a relocation and housing project must undergo four long stages with several time-consuming activities in each phase: (I) the pre-relocation or social-preparation phase, (II) relocation, (III) post-relocation and (IV) estate management. Phases I–III cover the project-development stage, whereas phase IV comprises monitoring and estate management (Ballesteros & Egana, 2012). Under the LGU's Guidebook for Local Housing Project/Program, the following are the procedures involved in the preparation and implementation of resettlement project (LGU Guidebook, p 20):

[16]A near-city or off-city relocation is usually approximately ≥ 40 km from the relocatees' original residence. This implies displacement of families from their current sources of livelihood, school, and other services to which they previously had access (ATDFW Oct 2013 position paper).

The step-by-step procedures to be observed are as follows:

1. Pre-Implementation
   - Project Identification
     (a) Resettlement planning
     (b) Baseline survey
     (c) Evaluation of sites
     (d) Selection of sites
2. Project Preparation
   - Community organization and participation
   - Plan preparation
   - Compensation scheme and eligibility criteria
3. Project Appraisal and Financing
   - Feasibility Study
   - Identification of funding
   - Identification of resource requirements
4. Detailed Engineering
   - Architectural and Engineering Designs
5. Implementation
   - Relocation
     (a) Site acquisition
     (b) Project construction
     (c) Land registration, titling, and payment of compensation
     (d) Award - Transfer arrangement
     (e) Strengthening community organization
   - Resettlement and Rehabilitation
     (a) Access to employment, training and credit
     (b) Shelter, infrastructure, social services
6. Post-Implementation
   - Estate management
   - Monitoring, appraisal, and evaluation

If stage 5 that implements the resettlement project and developing the relocation site is considered the most complex, time-consuming, and costly phase, stage 1 is said to be the longest, most tedious and critical process of all of the stages. It involves the creation of committees and sub-committees and lengthy consultations between the community of beneficiaries, the NHA, the LGU, and the developer of the relocation and housing. This set-up is indeed unresponsive to the PDR of disaster victims in case no relocation and housing units are available after an unexpected major calamity such as Typhoon Ketsana that hit the Philippines in 2009. This situation also happened to the victims of Typhoon Yolanda (Tropical Storm Haiyan) in Tacloban and Leyte. If one merely follows the usual procedures of building relocation sites according to existing laws, phase II for Yolanda victims is already time-consuming, complex, and costly, how much more if Phase I is accomplished first before a suitable relocation can be given to disaster victims? Delays can deprive disaster victims of the necessary PDR to urgently build their shattered lives. Thus, the national and local authorities were forced not only to build temporary shelters in Yolanda-stricken areas but also to simplify the usual bureaucratic and legal procedures and burdens to immediately rehabilitate thousands of homeless typhoon victims. The numerous procedures required by the UDHA which can create delays for resettlement projects, are not feasible for the immediate housing needs of disaster victims. However, under the present legal normative set-up, the Philippine government has no choice but to comply with the housing and relocation laws under UDHA in the absence of a more comprehensive PDR law that governs post-disaster management.

4.9 Plurality of Informal Normative Orders

The plurality of norms does not only exist in implementing the PDR of disaster victims under the PDRRMA but also in the intermediary and informal normative orders of the various intermediary groups that govern the inter-agency relations between government corporations as well as between private and public corporations involved in the government's system of relocation, housing, and post-disaster management. Before the PDRRMA was enacted in 2010, there were already *kalakaran*, or pre-existing formal and informal agreements and customs (*nakagawian*), between the various government and private organizations involved in the relocation and housing of the urban poor under the UDHA. These pre-existing ties and informal rules or customs constitute the semi-autonomous social fields that compete and alter the official law. The new PDRRMA did not sideline the modes of thinking and ways of doing things of people already in the low-cost housing and relocation of the government under UDHA. The old thinking or *habitus* of people of various public and private agencies working under UDHA was merely extended for the long-term PDR of disaster victims through groups or actors who just continued their relocation and housing projects for PDRRMA. "[E]ach member of an organization brings to it a habitus formed under specific past conditions, some of which will be shared with other members and some of which will differ from them substantially" (Emirbayer & Johnson, 2008, p. 4). Despite the new PDRRMA's legal provisions for PDR of disaster victims, the *habitus* and the various informal normative standards of implementers who work under UDHA continued to dominate the actual PDR of Typhoon Ketsana victims.

One of the most powerful determinants that influence the *habitus* and disposition of law enforcers or program implementers of PDRRMA in actualizing the official law are the dominant cultural normative values that shape informal rules and behavior of people under pre-existing UDHA cultural setting. Each organization or agency, whether public or private, has its own corporate culture, i.e., "customary and traditional way of thinking and doing things, and which each member must learn and at least partially accept, in order to be accepted into service in the firm."[17] In addition, government agencies have their own corporate cultures that shape the thinking and behavior of agency members. Although each organization or agency has its own organizational culture, people and groups can form alliances and networks bonded by some common cultural values that shape their common ties. Configurations of values may vary from one organization to the next. However, these values are generally taken from the dominant values of the larger society. In the Philippines, some of the popular and influential cultural values that shape social interaction as well as interpersonal and inter-organizational behavior and thus, in this case, affect the implementation of the PDRRMA include *pakikisama*, (camaraderie) *palakasan, (patronage), kanya-kanya (follow one's self or group),* and *utang-na-loob (debt-of-gratitude).* They also include other informal Filipino rules of reciprocity and personalism. Filipinos are personalistic in their ways: To them, loyalty, commitment, and *pakikisama* (camaraderie) are more important than nonpersonal traits such as efficiency and effectiveness (Franco, 1986, p. 36).

The Philippines has a strong legal tradition that originated from the Spanish era when the Spanish and Penal Codes were introduced into the country and reinforced by the American legal system during the American period with the transplantation of new commercial laws in the legal system. Despite this, customary laws and quasi-legal orders and other normative social fields, often shaped by dominant Filipino cultural values, compete and resist official laws resulting in unintended consequences different from the law's objectives. The PDR program for disaster victims under the new PDRRMA, such as that of Typhoon Ketsana victims in Rodriguez, Rizal, is no exception to this cultural and behavioral patterns in the Philippines. Despite the clear legal provisions of the Philippine procurement laws on the public bidding of government projects, such as low-cost housing, for instance, *palakasan* continues to determine winning bidders. Those who have strong connections with top housing officials often get lucrative government relocation contracts.

Crucial to the determination of the adequacy and quality of relocation sites and housing is the social field between the NHA and private developers, notwithstanding other informal social fields between top public officers and executives of developers. Under the NHA law, the NHA can purchase privately owned lands for housing and relocation purposes and develop them through joint ventures or other arrangements with private developers (Section 6). The NHA has accredited

[17] Definition of culture in organization. Available at http://www.zeepedia.com/read.php?the_organization_culture_adjustment_to_cultural_norms_psychological_contracts_organization_development&b=52&c=6.

developers who establish resettlements and build houses. Owing to a long history of doing business, these two parties have developed informal ties and reciprocity that tend to sidestep the original intentions of the law. This situation also happens in the social fields between private developers, the HULRB, and local licensing and regulatory units of LGUs. Through long acquaintances as well as social connections due to regular housing projects, they create informal ties and even illegal connections to facilitate the developer's needs in housing business such as land or subdivision permits, registrations, and clearances. In an interview with the Municipal Disaster Risk Reduction Management Office (MDRRMO), officials, for instance, were surprised why SRHP developers were given the necessary permits and certifications and were able to build the relocation despite the fact that the site is highly disaster-prone and inaccessible.

Through adoption of the pre-existing plurality of official laws and social fields in Philippine socialized housing under UDHA, the PDRRMA's post-disaster recovery for Typhoon Ketsana victims in Southville Rodriguez, Phase 8A, created more gaps and polycentricity in the enforcement of the PDRRMA's post-disaster recovery provisions, resulting in more negative unintended consequences, opposite to what the official law provides. The plurality of official laws as well as semi-autonomous social fields dominated by some negative cultural Filipino values between government and private organizations affect the implementation of the PDRRMA. Some of the alliances of the different public and private organizations involved in the program influenced by some dominant Filipino cultural values of *pakikisama (camaraderie), palakasan (patronage), kakilala (acquaintance), utang-na-loob (debt of gratitude)* and *kanya-kanya (following one's self or group)* affect the organizational coordination and implementation of the post-disaster-recovery for Typhoon Ketsana victims. Social acceptance and the cultural norm of reciprocity are important normative standards in Philippine culture and thus constitute a semi-autonomous social field in corporate behavior.

A plurality of social fields with normative standards that govern the PDR of disaster victims does not only exist in government laws, rules, and regulations but also in intermediary social fields that support and implement the PDRRMA's long-term PDR goals. The greater the number of laws, the greater the number of formal and informal social fields and their rules among the public and private groups that interpret and enforce them. Multiplicity and complexity of laws and legal and quasi-legal orders breed more discretion and conflict, which allows interest groups to exploit the amorphous internal and external legal systems. Tamanaha (2007) argued that coexisting or diverse bodies of law can result in competing claims of authority, conflicting demands of norms, diversity of styles and orientations in law enforcement, as well as conflict in coordination. "This state of conflict also creates opportunities for individuals and groups within society, who can opportunistically select from among coexisting legal authorities to advance their aims" (Tamanaha, 2007, p. 1). The diversity of laws and legal orders implies polycentric interpretation and implementation of the official normative standards and the involvement of more non-state actors and groups vying with state law enforcers in determining the actual implementation of the state's official law on PDR.

The Philippine socialized housing system is not only influenced by official law enforcers, donors, and legitimate business and bureaucrats but also by professional squatters, crime and housing syndicates, agents, power brokers as well as unscrupulous developers, politicians, and public officials who pursue their personal and corporate interests in competition with the goals of the official laws. The inadequacy of the legal provisions of the PDRRMA regarding post-disaster recovery, particularly the dynamics and procedures of providing relocation and adequate housing to disaster victims, has led to the adoption of a whole complex of laws, rules, procedures, and other norms on housing and relocation under UDHA that are not directly related to disaster recovery and thus pose conflict to the PDRRMA's PDR goals. The tedious and long process of providing relocation and housing also allowed the various pre-existing "semi-autonomous social fields" or pre-existing and informal contractual relations of the various organizations, groups or intermediaries to influence in the implementation of resettlement projects for the urban poor under the UDHA.

Before the enactment of the PDRRMA in 2010, there were already pre-existing social arrangements (*kalakaran*) and networks (*samahan*) among various housing groups that lobby and influence the implementation of the UDHA and other low-cost housing laws. Powerful corporate groups—such as NHA-accredited developers, suppliers, and contractors—have already established their own networks and informal rules in the relocation and low-cost housing program of the government and allied themselves with top government housing corporations in such a way as to form a complex chain that can compete with or resist the official state laws (Moore, 1973).

The PDRRMA explicitly aims to provide disaster victims with a better life in the relocation site after a disaster. However, government agencies and their private partners, which will perform these tasks, already have their own alliances and *nakagawian (customs)* on "how to make things done." For instance, the NHA, which is tasked to provide relocation and housing to disaster victims, has already established networks with LGUs, mayors, private organizations, and corporate developers. Corporate developers, in turn, also have established their own social fields with government coordinating and regulatory agencies such as the Department of Environment and Natural Resources (DENR) for environment clearances, Housing and Urban Land Regulatory Board (HULRB) for subdivision permits and clearances, Registry of Deeds (RD) for land titles, Housing and Development Coordinating Council (HUDCC) for policy coordination, and other private suppliers and constructions companies to facilitate other tasks.

4.10 Key Housing Intermediary Groups

Each social field consists of an intermediary network of two or more individuals or groups that serve as the bridge between the empirical reality and the official law. Not all fields oppose or resist the official laws though. Some—such as community organization or NGOs—lobby and work for the full implementation of the UDHA, PDRRMA, and other related laws. However, others work to deviate the purpose of the official law for their own advantage. They have their own hidden agenda and vested interests which conflict with the envisioned goals of the official law. The merging of legal and illegal networks through corruption in various organizations blurs the intended goals of the government's PDR program. The decriminalization of squatting under the UDHA is said to have created the unintended effect of creating various illegal intermediary groups with their illegal informal rules, which undermine the goals of the official state laws. The UDHA or R.A 7279 through R. A. 8368, which repealed P.D. 772, penalizes squatting with imprisonment. Under Section 1:

> Any person who, with the use of force, intimidation or threat, or taking advantage of the absence or tolerance of the landowner, succeeds in occupying or possessing the property of the latter against his will for residential commercial or any other purposes, shall be punished by an imprisonment ranging from six months to one year or a fine of not less than one thousand nor more than five thousand pesos at the discretion of the court, with subsidiary imprisonment in case of insolvency.

R.A. 7279 or UDHA, recognizing that squatting is a social problem caused by poverty and lack of urban-land reform rather than a problem of criminality, removed the criminal liability of poor people occupying other peoples' land. This law prosecutes professional squatters but explicitly excludes from the definition individuals or groups who simply rent land and housing from professional squatters or squatting syndicates. However, there are no reliable criteria and monitoring tools to distinguish the legitimate urban poor from illegal groups pretending to be poor such as the professional squatters. As a result, illegal groups emerged and took advantage of the government housing program contrary to UDHA's pro-poor legal provisions. The PDRRMA inherited these pre-existing illegal groups when it relied on the UDHA and its legal infrastructure for the long-term PDR of disaster victims. Those who illegally occupy private or public land, called "squatters," can be classified into three types of groups: the poor, professional squatters, and squatting syndicates.

4.10.1 Professional Squatters

One of the most powerful and illegal intermediary groups that oppose the goals of UDHA and compete with government law enforcers in implementing the official law are the professional squatters. The UDHA defines "professional squatters" as individuals or groups who occupy lands without the express consent of the land-owner and who have sufficient income for legitimate housing. The term also applies to persons who have previously been awarded home lots or housing units by the government but who sold, leased, or transferred the same to settle illegally in the same place or in another urban area as well as and non–bona fide occupants and intruders of lands reserved for socialized housing. The term does not apply to individuals or groups who simply rent land and housing from professional squatters or squatting syndicates (Section 3 [m]).

Although the UDHA labeled beneficiaries who sold their home lots or housing units as professional squatters, not all beneficiaries who sold their awarded homes in the resettlement area can be considered professional squatters. In SRHP, Phase 8A, for instance, some Typhoon Ketsana victims sold their housing units because of poverty and expenses of commuting daily to work from the relocation area in the remote Barangay of San Isidro to some urban centers of Metro Manila. The daily fare alone can eat up almost half of their daily wage or income.

Professional squatters may vary in form and constitution. However, the most common form of these squatting groups, according to key informants, are usually composed of relatives who are not really poor and homeless but would only want to avail themselves of the government's housing program for profit. These people usually reside in squatter communities and occupy a publicly or privately owned land until they become qualified to apply as beneficiaries in the government's low-cost housing program. After being awarded with housing units, they move again to another "squatters' community" and continue their illegal activities. They either lease or sell to other people the housing units awarded to them by the NHA. Some of these people have connections with local politicians and officials who can assist and protect them in their illegal activities.

The professional squatters are people who see some loopholes in the government low-cost housing system to turn the government's relocation program into a profitable business. These people are not actually destitute or poor but live in slum areas or public lands occupied by informal settlers with the goal of gaining a housing unit once the NHA transforms them into low-cost housing sites. Once awarded the units, they transfer to another area to gain another housing unit. One key informant who is a professional squatter and gainfully working in a super-market confessed that he has five different housing units in various government low-cost subdivisions, which he rents to other people to earn more income. He was able to acquire these units by residing in various squatter colonies and availing himself of the NHA housing program.

4.10.2 Squatting Syndicates

There is an overlap in the definition of professional squatters and squatting syn-dicates. Both are engaged in the business of squatting. The major difference is that squatting syndicates are professional squatters who constitute the power elite of a squatting organization and control one or more squatting colonies. Neal Cruz, in his newspaper column, aptly describes the role of a squatting syndicate in a squatter colony:

> There is a syndicate behind every squatter colony. When a colony is starting, outsiders will arrive to organize the squatters into an association, assuring them of protection from eviction. They collect monthly membership fees of P500 to P1000 from each family, threatening eviction from the colony if anyone protests. There is no accounting of these membership dues, which just disappear into the pockets of syndicate members (Neal Cruz, PDI 23 July 2013). The syndicates have lawyers and enforcement teams composed of former soldiers and policemen, some of whom still carry their service firearms.[18]

The UDHA defines "squatting syndicates" as "groups of persons engaged in the business of squatter housing for profit or gain" (Section 3, [s]). They are informal and illegal organizations that covertly coordinate the activities of professional squatters (Ragragio, p. 7). There is no specific Philippine statute that penalizes squatting syndicates. There is no statement in penal-statute books in the country stating that a person, corporation, association, or organization shall be punished for engaging in a squatting business and selling rights over a parcel of land not belonging to them.[19]

The squatting syndicates are similar to professional squatters in a sense both groups avail themselves of the government's housing and resettlement programs illicitly for personal gain. Both are disqualified by law to become beneficiaries of the low-cost housing of the government intended only for poor people with no real property. Squatting syndicates operate more like an illegal real-estate enterprise. According to informants, these groups have money to bribe unscrupulous NHA officials to be able to acquire housing units, which they then sell to others for profit. These people also hire poor people to occupy the housing units. The HPDO and Rodriguez law enforcers have already intercepted such people who were about to enter SRHP in order to illegally occupy some units in the relocation site upon the orders of housing syndicates who have connections in the NHA.

Another form of squatting syndicates are the more organized groups of "land grabbers" who have illegal titles, occupy government-owned lands, and allow their own people to squat the land. Eleven suspects, for instance, who belong to this type of syndicates were arrested in Quezon City. Using a Certificate of Land Occupancy issued by a certain Allan Bustilo, alleged attorney-in-fact of the Doña Lourdes Rodriguez Yaneza Estate. This atorney-in-fact shows ownership by virtue of *Titulo de Terrenos* Royal Decree OLT 01-4 with TCT No. 38400. The suspects posed as military reservists and occupied a vacant lot, which owned by the government and held in trust to the NHA inside the National Government Center Housing Project West MRB Compound in Quezon City, and duped poor people to

---

[18]http://opinion.inquirer.net/55897/squatting-syndicates-victimizing-real-squatters.

[19]http://www.philstar.com/headlines/2015/04/10/1442360/anti-professional-squatting-drive-reinforced.

join their organization with a promise that they can own a house and lot worth PHP50,000.[20]

Another strategy according to informants is that a squatting syndicate occupies government land with their illegal occupants. When the government converts the location into a low-cost housing project, this syndicate illegally acquires housing units and either sell or lease them for profit. Then people then transfer to another targeted public land and wait for the occupied land to be converted by the NHA into a low-cost housing project.

Finally, squatting syndicates can illegally acquire housing units in the government's relocation sites directly from the NHA through their contacts in the agency. Then they hire or contract people to occupy the units in the relocation site. The case of a group of illegal relocatees riding in dump trucks with an NHA banner bound for Southville Rodriguez which was intercepted by the HPDO and local law enforcers, is an example of this illegal strategy of squatting syndicates to take advantage of the government's resettlement program.

4.10.3 Political Patrons

The political patrons are incumbent and non-incumbent officials in the national and local governments who have strong political connections with top government officers with a network of constituents, relatives, and friends. In the local level, local politicians in the LGUs can constitute a network of political patrons who control the distribution of housing units in the relocation site. These people form a network primarily based on political affiliation and patronage. In a local government such as Rodriguez municipality, this scheme comprises the top three politicians in the province—the governor, the district congressmen, and the municipal mayor—and their barangay captains and supporters to form a strong network that controls the distribution of housing units in the relocation site. These politicians usually belong to one political party with the provincial governor as the informal leader of this group. In Rizal province, a mayor or congressman can hardly win elections without the support of the provincial governor, who comes from a political dynasty that has controlled the provincial and local politics for a long time. On paper, the UDHA requires that beneficiaries must be poor, homeless without real property, and disaster victims as demanded by the PDRRMA. However, in practice, access to the government's resettlement and housing programs is determined by the people who are directly or indirectly affiliated with the political patrons of the province.

Outside Rodriguez, such as Metro Manila, people can also avail of the housing units at SRHP through the network of mayor, congressmen, barangay captains, and their supporters under one city or municipality that forms a powerful group who determines who among their constituents can immediately receive housing benefits.

[20]http://www.philstar.com/metro/453677/quezon-city-squatting-syndicate-neutralized.

4.10.4 Housing Brokers and Agents

The lack of knowledge of many buyers and beneficiaries on real-estate laws and procedures, as well as the bureaucratic and complex real-estate system in the Philippines, has encouraged the active participation of real-estate agents in processing housing transactions, including low-cost housing, for the urban poor and disaster victims. According to informants, there are two types of housing brokers and agents that can influence the distribution of housing units in the relocation: (1) legitimate housing brokers and agents of the developers and (2) fly-by-night informal agents in the relocation area.

Some buyers and beneficiaries deal with brokers and agents without verifying the authenticity of documents and transactions with the NHA and Land Registration Authority (LRA). They also approach informal agents of "fixers" to expedite the transaction. Sometimes, they pay the reservation and other fees. The "fly-by-night" agents are usually recommended by *kakilala* (acquaintance). However, because poor buyers do not have the means and knowledge on how to verify the title and authenticity of the transaction, double-sales often happened, according to informants. In such case, the units promised by agents are already occupied by or titled to other claimants. At times, buyers or beneficiaries may not immediately aware that others are already occupying the housing units. Because housing units are bare, unfinished, and unfurnished, they delay their occupancy until they have money to fully develop their units. Once they try to move in, they often discover that other people have already owned and occupied their units.

Some of the informal agents do not have license and expertise in real-estate business. According to informants, a few has just learned the basic steps and requirements in applying for low-cost housing. Although they only learned the basic steps, they immediately became informal agents to housing applicants.

Many of these agents are friends and acquaintances of housing beneficiaries who illegally sell housing units. They are also residents in the relocation area. According to informants, the sale of housing units in SRHP is popular. It is common that some people who act as agents approach friends or visitors in the relocation site and invite them to buy the rights of some housing units. Some of these agents may be relatives or friends of the beneficiaries who also receive some commissions if a sale is consummated. To account for the agent's commission, the contract price of the rights of the house and lot is usually increased. One informant who sold her housing unit revealed that she directly sold her rights directly to a buyer to avoid increasing the selling price. During the researcher's to the relocation site, he was offered by these agents to buy some housing units in the area that cost approximately PHP 50,000–100,000 (US$ 1,000–2,000).

Under this illegal arrangement, the buyer of the rights of the housing unit continues the monthly amortization until the house and lot are fully paid in the NHA. Then the seller executes an absolute deed of sale transferring ownership to the buyer. According to the official law, it is illegal for the beneficiary to sell his or her rights to others. He or she would be considered a professional squatter and could no longer avail of any housing benefit from the government in the future. The true intention of the PDRRMA is thus frustrated. Most of the buyers are non-victims of disasters, and some are professional squatters who would also sell or rent the units to gain profit.

The existence of relocation-housing agents is an unintended effect of the government's policy of building relocation sites in remote areas and far from urban jobs and livelihood activities. Because of the cost of the fare, daily expenses, and inconvenience of traveling long distances just to work in the informal sector, many beneficiaries sell their housing units with the help of informal agents and return to their original residence or live as informal settlers near to their former place of work or trade.

4.10.5 Major Cultural Informal Normative Systems

Normative systems operate in networks. Instrumental Filipino values often interact and overlap with one another to form a normative system and compete with official laws and regulations on housing and resettlement projects. In particular, the awarding of housing contracts and the distribution of housing units in the post-disaster relocation in the Philippines often follow these normative systems and sideline the official norms.

4.10.5.1 The *Palakasan* Normative System

Some major informal norms that compete with the implementation of the official laws on relocation and housing, especially with regard to access, are (1) *palakasan* (patronage) and (2) the *Padrino* (sponsorship) system. These two interrelated systems are based on acquaintance (*kakilala*) rather on merits as demanded by the official law. They are in consonance with the personalistic character of the Filipino culture. A secondaryanalysis by Abad (2006) on social capital showed that Filipinos tend to approach their families, relatives, and friends for a favor and not the insignificant others outside of these networks. The more the requesting party is socially known and connected to the person or party dispensing the favor or government benefit, the greater the probability of receiving it.

4.10.5.2 The *Padrino* (Sponsorship) Normative System

The *Padrino* system in the Filipino culture and politics is a cultural normative system where one gains favor, promotion, or political appointment through family affiliation (nepotism) or friendship (cronyism), as opposed to one's merit."[21] It is based on ritual or fictive kinship and created through sponsorship in Catholic baptism, confirmation, or matrimony. *Padrino* is the Spanish equivalent to the Filipino words *ninong/ninang* (godfather/godmother) and *compadre/kumara* (co-father/co-mother) and refers to the patron who may be a family, school, or community leader. This normative system is built on trust and based on patron–client relationship, which allows actors to construct meaning, provide access to power and influence, and create opportunity for both the patron and the client (Normare and Erbe (Eds.) (2014), p. 208)

Under *palakasan* and the *Padrino* system, a person or group who wants to get benefits or favors from the government must have a "backer" or backers (*Padrino* or sponsors), i.e., a person or group of people who are politically powerful and influential (*malakas*) in government bureaucracy either as a relative (*kamag-anak*), friend (*kaibigan*), or godparent (*ninong/ninang*), a former classmate (*kaklase*), and so on. In *palakasan* and the *Padrino* system, those who want to avail of government benefit or position must personally know an influential group or individual politician who is related to the beneficiary either by consanguinity, affinity, friendship, fictive relationship, or other form of affiliation.

Kamag-anak, or kinship ties, is an important informal norm that influences the accessibility of the government's housing projects in the Philippines. And kinship ties in the Philippines, being a Catholic country, are based not only limited to consanguinity or affinity but also include *compadrazgo* or ceremonial or fictive kinship ties. Informants revealed that relatives of some HPDO officials who were neither poor nor disaster victims were awarded housing units in the relocation. One of the developers of Southville Rodriguez, as reported by (Philippine Center for Investigative Journalism) PCIJ, is the brother-in-law of the executive secretary of a former Philippine President.

Being a former classmate (*kaklase*) and friend (*kaibigan*) is also an important informal norm that downplays the power of the official law. Being friends and former classmates also influences access to government housing benefits. A partner of the developer of Erap City, as reported by PCIJ, was a former high-school classmate of President Erap Estrada.

The degree of power one gets from the *palakasan* system depends on the power and influence of the *Padrino* or sponsor. The greater the position of the sponsor in the political hierarchy, the more *malakas* (powerful) the person or group with regard to housing benefits. It depends on who you are clinging to (*depende kung sino ang kinakapitan*). Thus, an applicant with the incumbent governor as the sponsor has more chances of getting a housing unit than one who has the barangay captain as the backer. The informal rule of nepotism or kinship ties is also important in the *palakasan* system. An applicant who is a relative of the sponsor has more chances than a non-relative or a mere political supporter. Fictive kinship ties—such as having a politician acting as godfather (*ninong) or* godmother (*ninang*) in a

[21] http://www.encyclo.co.uk/meaning-of-Padrino%20System.

wedding or baptism also counts in *palakasan or the Padrino* system. Politicians in the Philippines are known to act as sponsors in weddings and baptisms to gain more votes.

4.10.5.3 The *Utang-Na-Loob* (Debt of Gratitude) Normative System

Mary Racelis-Hollnsteiner, who pioneered a study of the well-known Filipino value of *utang-na-loob,* defines it as "that principle of behavior wherein every service received, solicited or not, demands a return, the nature and proportion of the return determined by the relative statuses of the parties involved and the kind of exchange at issue" (1973, pp. 69–91). It rests in between two extremes in a continuum: *hiya* at one end, where one who received a favor does not reciprocate or return a favor received from the other, and the purely businesslike contractual reciprocity demanded by official law.

The awarding of housing units in the relocation follows this informal norm of *utang-na-loob*. The incumbent local politicians return the favor or votes they received from people who requested for houses in the relocation houses. This is a major reason why the number of houses allotted by the NHA to the local government is further sub-divided among the top incumbent officials of the province. The distribution of the housing units is largely determined by *utang-na-loob*. Those who supported, campaigned, or voted for these incumbent officials during the last election have a greater chance of being awarded housing units in the relocation area, even though they are not qualified as beneficiaries under the UDHA and PDRRMA. That is why the value of *utang-na-loob* can lead to *pakiusap* (negotiation), nepotism and "cronyism," in the distribution of government benefits (Gorospe, 1988, p. 68).

According to HPDO informants, housings units at the SRHP allotted by the NHA to the local government are divided equally among the top politicians of the province for their disposal: one third of the total number of houses to the governor, one third to the congressman, and one third to the municipal mayor. Although the official law of the UDHA sets the qualification for the housing beneficiaries, our informants believed that the distribution and awarding follow the informal norm of *utang-na-loob.* Applicants who are strongly identified with the incumbent officials as backers, supporters, campaigners, or voters are more likely to be approved than those who are not politically identified with these three top officials in the province where the relocation site is situated. This has been confirmed by some beneficiaries in the Southville Rodriguez, Phase 8A area, who admitted that the approval of their housing applications was expedited because of their connections with some incumbent officials whom they supported during the last elections. Two beneficiaries, however, confessed that it took them 6 months to receive the approval of their applications because of their lack of *padulas (grease money)* or political backers to their applications.

4.10.5.4 The *Sakop, Kanya-Kanya*, or *Grupo-Grupo* Syndrome

Some social scientists have categorized culture into individualist and collectivist. Asian and Oriental cultures, with their emphasis on the community rather than the individual, are generally considered "collectivist," whereas Western and American cultures, which stress the power of the individual over that of the community, are generally labeled "individualist." Under this categorization, the Philippines can generally be classified as having a collectivist culture where the group and community are given more importance than the individual in decision-making. Although official disaster laws emphasize welfare and equality in the distribution of disaster assistance, informal cultural norms, which give preference to the in-group rather than the out-group and individual welfare, are often followed in social practice. The Philippines as a collectivist culture puts more premium on the welfare of the *sakop* (in-group) rather than the public welfare and equal access to public goods and services. In relation to Philippine politics, the concept of *sakop* is connected with political patronage. The head of the *sakop* is often a political patron.

With this cultural arrangement, the distribution of housing units in the relocation area in the Philippines is often based on informal rules of *sakop* and political patronage rather than on merit, and the *sakop* or *grupo-grupo* (group oriented) syndrome dominates the system of relocating disaster victims in the resettlement. The informal norm of *sakop* can breed *kanya-kanya,* which is a self-serving attitude that generates a feeling of envy and competitiveness toward others. This not only denotes individualism but also exclusiveness of groups–that one group is different from or in competition with other groups. Anyone who is not a member of the *sakop* or group cannot avail of the benefits it receives from the government. Thus, disaster victims who belong to another group or *sakop*—or who are not affiliated with political patrons by kinship, political party, or any form of the social network of one group or *sakop* who received housing units—would have a difficult time getting their housing applications approved. In SRHP, some Metro Manila mayors at times coordinate directly with the NHA and bring their own constituents directly to the relocation area without the knowledge and approval of the local municipal government. This contravenes the official law, which requires coordination with the local government where the relocation is located.

4.10.5.5 The *Padulas* and *Lagay* Normative Systems

Monetary and bureaucratic corruptions sidestep the legally prescribed rules. As already mentioned, corruption is the "colonization of social relations in which two or more actors undertake an exchange relation by way of a successful transfer of the steering media of money or power, thereby sidestepping the legally prescribed procedure to regulate the relation" (Deflem, 1995, p. 1). Corruption follows the cultural norm of reciprocity: "You scratch my back, and I'll scratch yours." Corruption can be either monetary or bureaucratic. Monetary corruption is the exchange relation between two or more actors that is performed by way of a transfer of money, whereas bureaucratic corruption is carried out by a transfer of power.

The informal rule of *padulas* (grease money) is often used by people in the Philippines who want to expedite their requests or transactions with government agencies. A number of *padulas* depends on the type and amount of the transaction involved. This is a form of *lagay* (bribe), which can be either an amount of money or an in-kind, such as material gifts, in order to get fast approval of one's request for permits, clearance, or registration from government-housing regulatory bodies. In the Registry of Deeds, for instance, one lawyer–informant revealed that a minimum of PHP 500 *lagay* is usually given to every title signed by the Register of Deeds (RD). Thus, if one developer hands in 100 titles for signature in one transaction, the RD receives PHP 50,000 amount of *lagay*. In obtaining an environmental clearance or ECC, a huge *lagay* is needed especially if the location is unfit for subdivision or housing project as required by environmental laws. Some key informants of the local government of Rodriguez believe that *lagay* and bureaucratic corruption are the main reasons why developers of SRHP were allowed by government regulatory bodies to expand the Erap City and construct the SRHP despite the risks and geo-hazards of the location and the classification of the site as a disaster-prone by the Municipal Disaster Risk Reduction Management Council (MDRRMC).

The giving of *lagay* is also a common norm for housing syndicates who want to control housing units in the relocation site. Key informants revealed that members of the local housing office are usually given *lagay* by people, especially by members of the housing syndicates, who want their applications immediately approved by the office. One official of the HPDO recalled an experience where an applicant followed him in his car after office hours and handed him an envelope containing PHP 5000 (around US$100) in exchange for approval of a housing application in the relocation site. The giving of *lagay* is a common practice in local government housing offices. This is one reason why the mayor of Rodriguez appointed a religious pastor to head the HPDO to stop or minimize the informal rule of giving *lagay* in order to obtain housing units in the relocation site despite being disqualified by law.

The informal rule of *tongpats (price padding)* is another form of monetary and bureaucratic corruption. *Tongpats* is a Filipino slang and inversion of the word *patong* or padding of prices. It implies that the original government contract price is padded or increased considerably in order to create a commission (kickback) to whoever is brokering the deal.[22] This term became popular when the whistleblower Rodolfo "Jun" Lozada, Jr. of the botched ZTE deal revealed in the media that the contract price of the railway project during the Arroyo administration was padded or with *tongpats* to provide kickback for powerful individuals who orchestrated the public bidding. Informants believe that like any government bidding for low-cost housing, projects at the NHA are also tainted with corruption.

---

[22]http://pcij.org/blog/2008/03/12/the-language-and-manner-of-doing-shady-business.

Overall, this multiplicity of laws and semi-autonomous fields and agencies tasked to interpret and actualize the PDRRMA's post-disaster recovery goals for calamity victims has resulted in many negative unintended effects. Thus, some legal scholars believe that the study of law must not only focus on how emergent norms within a given social field affect state laws but also on how state laws impinge on social fields that already have customs and informal rules. Today, legislation that is primarily intended to direct change in society often resulted in unintended consequences because existing social relations tend to be stronger than new laws (Moore 1973, p. 58). The succeeding chapters will illustrate how pre-existing social relations and their major informal norms based on some major Filipino values affected the implementation of the PDRRMA and UDHA in the relocation of Typhoon Ketsana victims in SRHP Phase 8A in Rodriguez, Rizal.

4.10.6 Consequences of Normative Pluralism

The multiplicity of laws and informal normative orders vying to interpret and implement the official PDRRMA on PDR has also affected the coordination and cooperation of various organizations involved in enforcing the government's PDR project under PDRRMA, which further created more unintended consequences to the project. Public and private organizations—as well as informal groups, power cliques, and foreign donors—tend to have their own established formal and informal rules of engagement in their own domains. Among government agencies, there are operations at cross-purposes; thus, conflict and inconsistencies within and between relevant agencies can occur, which hamper the legal implementation (Castellani, 1992). Large-scale disasters such as Typhoon Ketsana usually involve multiple organizations and agencies having the capacity to coordinate their efforts in the management of disasters (Drabek, 1985; Drabek & McEntire, 2002; Quarantelli, 2000). However each organization or agency has its own mandate, pre-existing social networks, and discretionary powers as to when and how it coordinates with another agency. Thus, when new disaster-related tasks are to be performed, questions almost inevitably arise about which organizations have the authority to assume them (Quarantelli, 1988). In the Philippines, the coordination of government-initiated post-disaster housing projects largely depends on pre-existing networks and ties, primarily influenced by cultural normative orders with their dominant Filipino instrumental cultural values as informal norms. The Filipino values of *kakilala*, *utang-na-loob, palakasan, Padrino*, or *sakop*, which emphasize the in-group rather than the out-group coordination between private and public agencies implementing the PDR, can create difficulties in implementing the legal norms of the PDR projects.

When the PDR provisions of disaster law are broad and brief, organizational authority conflicts often arise—albeit the pre-existing organizational ties tend to facilitate new coordination efforts. Thus, national and local government units often clash in interpreting and determining the provisions of the official law. In this case, discretion enters the picture, and the most influential actor or group tends to prevail over weaker ones in disaster situations. Discretion in decision-making refers to the ability of a public officer to make a choice among possible courses of action and inaction whether legal, illegal, or with questionable legality within the effective limits of the office (Pepinsky, 1978). In relation to legislation, "discretion" refers to that substantive space in the law that enables law enforcers to implement the statute they think is a best fit for the situation. This exercise of discretion is not independent of ongoing social arrangements of people competing with the state law. Personal discretion is influenced by various social fields where the person or group is situated.

The failure of the government to enact a comprehensive *Magna Carta* on PDR that (1) details the post-disaster relocation of disaster victims, (2) appoints a separate group of disaster managers different from incumbent public officers, and (3) creates a new national post-disaster management agency that separate from the resettlement of the urban poor unaffected by natural disasters has led to a complex legal pluralism and discordant normative orders to implement the official legal PDRRMA. In legislation, the scope of the substantive law determines the discretionary space of coordinating agencies as well as the bureaucrats who implement the law. Although lengthier and more specific legal provisions limit the discretion of law enforcers, shorter and vaguer ones are said to create more discretion and loopholes, which can result in more unintended consequences in the implementation process (Evans (Ed), 1978). Furthermore, vaguer legislations can allow pre-existing legal order and other normative social arrangements to influence the outcome of the official law. The legal gaps can ultimately result in loopholes in the implementation stage that allow hegemonic groups in the PDR project, as well as their informal normative systems based on some dominant negative Filipino values, to influence the outcome of the project.

4.11 Summary

This chapter has described legal and normative pluralism in terms of laws, social fields, and major informal rules based on some Filipino values that influenced the interpretation and implementation of the PDRRMA's post-disaster recovery provision under UDHA. The PDRRMA's lack of comprehensive provisions on how the relocation and housing program for the disaster victims should proceed has led to the adoption of UDHA and its entire array of laws on housing and relocation of the urban poor at both the national and local levels. The normative pluralism that surrounded the implementation of PDRRMA's post-disaster recovery did not only exist in terms of housing laws and government regulations and rules but also in terms of social fields and intermediary groups with their informal rules based on some dominant Filipino cultural values. By adopting the UDHA's relocation and housing provisions, the PDRRMA inherited the UDHA's pre-existing problems such as the existence of illegal and corrupt intermediary groups and informal rules that subverted the UDHA and PDRRMA's legal goals on relocation. The decriminalization of squatting by the UDHA had inadvertently resulted in the proliferation of major illegal intermediary groups such as squatting syndicates, professional squatters, political patrons, or illegal housing agents. The official housing and relocation laws were in competition with some informal rules based on Filipino values of reciprocity such as *palakasan* and the *Padrino system*, *utang-na-loob*, *padulas*, *lagay*, *tongpats*, and other related informal norms resulting in numerous unintended negative consequences for the post-disaster resettlement of Typhoon Ketsana victims in SRHP in the Province of Rizal.

References

Ballesteros, M. F. (2000). Land Use Planning in Metro Manila and the Urban Fringe: Implications on Land and Real Estate Market. The PIDS Discussion Paper Series No. 2000–20. http://citeseerx.ist.psu.edu/viewdoc/download?doi=10.1.1.202.2710&rep=rep1&type=pdf

Ballesteros, M. F. (2002). *Rethinking Institutional Reforms in the Philippine Housing Sector*. Philippine Institute for Development Studies (PIDS). Perspective Paper Series No. 11. https://dirp3.pids.gov.ph/ris/books/pidsbk05-ppshousing.pdf

Ballesteros, M. & Egana, J. (2012). Efficiency and effectiveness review of the National Housing Authority (NHA) resettlement program. http://www.dbm.gov.ph/62C97C27-20B5-4CB0-9BD2-8286EDF12907/FinalDownload/DownloadId-802ACD5E5502BED20A5F66A855713/62C97C27-20B5-4CB0-9BD2-8286EDF12907/wp-content/uploads/DBM%20Publications/FPB/ZBB-2012/e.pdf. Accessed March 4, 2015.

Castellani, P. (1992). "Closing Institutions in New York State: Implementation and Management Lessons." *Journal of Policy Analysis and Management, 11*(4):593–611.

Congressional Policy and Budget Research Department (CPBRD). (2011). Addressing policy gaps in the land administration system. CPBRD Policy Brief. No. 2011–4. May 2011.

COA Report. (2004). *Sectoral Performance Audit Report on the Housing Program of the Government*. CYs 2001–2004. http://www.coa.gov.ph/index.php/sectoral-performance-audit/category/244-housing-program-of-the-government?download=701:national-housing-authority-ms-report-no-2004-02a

Deflem, M. (1995). Corruption, law, and justice: A conceptual clarification. *Journal of Criminal Justice, 23*(3): 243–258.

Drabek, T. E. (1985). Disaster Management: The Human Factor. Emmitsburg, MD: National Emergency Training Center, Federal Agency Management Center.

Emirbayer, M., & Johnson, V. (2008). Bourdieu and organizational analysis. *Theory and Society, 37*(1), 1–44.

Franco, E. A. (1986). *Pinoy Management*. Metro Manila: National Bookstore.

Hollnsteiner, M. (1960). *The Dynamics of Power in a Philippine Municipality*. Community Development Research Council: University of the Philippines.

Hollsteiner, M. R. (1973). Reciprocity in the Lowland Philippines. In F. Lynch & A. de Guzman II (Eds.), *Four readings on Philippines* values (pp. 69–91). Quezon City: Ateneo de Manila Institute of Philippine Culture.

IFRC. (2015). Impact of the regulatory barriers to providing emergency and transitional shelter after a disaster: Country case study: Haiti. http://www.ifrc.org/PageFile.

Moore, S. F. (1973). Law and social Change: The Semi-autonomous social field as an appropriate subject of study. *Law and Society*, *7*, 719–746. http://keur.eldoc.ub.rug.nl/FILES/Wetenschapterschapters/2/730/730. Accessed February 10, 2013.

Normare, A., & Erbe, N. (Eds.). (2014). *Collective efficacy: Interdisciplinary perspectives on international leadership*. United Kingdom: Emerald Group Publishing Limited.
Pepinsky, H. (1978). Discretion and Crime Legislation. In M. Evans (ed). *Discretion and Control*. London: Sage.
Quarantelli, E. L. (1988). Diaster crisis management: A summary of research findings. *Journal of Management Studies, 25*: 373–385. doi: 10.1111/j.1467-6486.1988.tb00043.x
Quarantelli, E. L. (2000). Disaster research. In: Borgatta HEF and Montgomery RJV (eds) Encyclopedia of Sociology, 2nd edn. New York: Macmillan, 681–688.
Smart, C. (1986). *Feminism and the Power of Law*. London: Sage Publication.
Tamanaha, B. (2007). Understanding legal pluralism: The past to present, local to global. *Sydney Law Review, 29*. St. John's Legal Studies Research Paper No. 07-0080. Available at SSRN: http://ssrn.com/abstract=1010105
Valte, M. (2013). A House for Every Filipino: A Pipe Dream in the Land of Promises? http://www.ombudsman.gov.ph/UNDP4/wp-content/uploads/.../Chap6.pdf
World Bank Group (2014). Doing Business Economy Profile 2015: Philippines. Washington, DC: World Bank. https://openknowledge.worldbank.org/handle/10986/21179. License: CC BY 3.0 IGO.
World Bank Group (2016). Doing Business 2016: Measuring Regulatory Quality and Efficiency. Washington, DC: World Bank. http://www.doingbusiness.org/~/media/WBG/DoingBusiness/Documents/Annual-Reports/English/DB16-Full-Report.pdf

Chapter 5
Suitability and Accessibility of the Relocation Site

Abstract This chapter analyzed closely the suitability and accessibility of the Southville Rodriguez Housing Project (SRHP), a government relocation site for Typhoon Ketsana victims in the Philippines, *vis-à-vis* the legal provisions of international conventions on post-disaster recovery of displaced persons and the country's Philippine Disaster Risk Reduction Act of 2010 (PDRRMA) on post-disaster management of disaster victims. It examined the official laws and standards regarding the selection and accessibility of relocation sites for disaster victims under the "build back better" (BBB) principle and described the extent of conformity of the current resettlement of the Typhoon Ketsana victims to these standards. Owing to corruption and undue benefits given by the government to its accredited private developers of resettlements, the conditions of the relocation site of Typhoon Ketsana victims in SRHP largely deviated from the international standards and provisions of PDRRMA regarding the suitability and accessibility of relocation sites for disaster victims. The SRHP is not only remote, dangerous, and disaster-prone, but it is also physically and socially inaccessible to poor disaster victims.

Keywords Disaster management · Resettlement · Post-disaster housing · Suitability · Accessibility · Corruption

5.1 Introduction

One of the essential components of a holistic and integrated long-term post-disaster recovery (PDR) program for disaster victims is adequate housing and relocation. The Philippine Disaster Risk Reduction Management Act of 2010 (PDRRMA), the Urban Development and Housing Act of 1992 (UDHA), and the United Nations' guidelines for the resettlement of internally displaced persons require that the shelter and relocation for disaster victims must be suitable for their recovery process. They also prescribe that the relocation must be accessible to all, particularly for marginalized persons. The criteria of suitability, on one hand, stipulates that the housing and relocation must be safe and appropriate to the PDR process of the

V.O. Ballano, *Law, Normative Pluralism, and Post-Disaster Recovery*,
DOI 10.1007/978-981-10-5074-9_5

disaster victims. It implies that the housing units and relocation site must be built in safer grounds and disaster-free, thus ensuring the avoidance of vulnerability that disaster victims have already experienced. It also requires that the location of the socialized housing and resettlement projects are proximate to areas where employment opportunities are accessible. The criteria of accessibility, on the other hand, requires that the low-cost housing project is available to all granted without discrimination, especially to marginalized persons, and must be physically and financially affordable to the poor. This chapter evaluates the suitability and accessibility of the Southville Rodriguez Housing Project (SRHP) relocation site, particularly the appropriateness of the location and housing units to Typhoon Ketsana victims' economic and social needs. Lastly, it suggests some sociological explanations on why legal and normative pluralism has resulted in more negative consequences that are contrary to the legal provisions of the PDRRMA on the PDR of disaster victims.

5.2 Determining Suitability of the Relocation Site

The PDRRMA has not explicitly indicated the criteria for determining the suitability of a relocation site for the PDR needs of disaster victims. This lack of specific legal provisions of the law on relocation and housing for disaster victims has led to the adoption of existing state laws such as the Urban Housing and Development Act (UDHA) of 1992 and other affiliated housing laws and their implementing rules to determine the suitability of the location site for the PDR of disaster victims. These laws and implementing rules are not designed for the PDR of disaster victims but for the relocation and housing of the urban poor who live in danger zones and are victims of demolitions in cities or municipalities.

According to the UDHA and its LGU guidebook for socialized housing, the first phase of the resettlement and housing of disaster victims is the social preparation of the site. This phase is said to be the most important and crucial phase of the resettlement process because it involves various groups and comprises various activities that are crucial in determining the suitability of the resettlement area. Although official laws on disaster management of the United Nations (UN) and the PDRRMA point to the national government as being responsible for providing suitable housing and relocation sites to disaster victims, its final realization would depend on many factors, such as the implementation of local laws on relocation and housing and the discretion of groups and government agencies involved in the social preparation of the relocation project.

5.3 Laws Governing the Suitability of Relocation and Housing

Before the National Housing Authority (NHA) and its accredited private developer can start building a relocation site and construct low-cost housing for disaster victims, the basic standards on relocation and socialized housing set by laws and administered by the Housing and Urban Land Regulatory Board (HULB) must first be followed before a permit or registration is issued to the developer.

Concerning physical suitability, Presidential Decree (P.D.) No. 957 and its revised implementing rules and regulations in 2009 under site criteria require the following:

1. Subdivision projects shall be located within suitable sites for housing and outside hazard-prone areas and protection areas as provided for by pertinent laws. Critical areas (e.g., areas subject to flooding, landslides and those with unstable soil) must be avoided.
2. The site shall be stable enough to accommodate foundation load without excessive earthmoving, grading or cutting and filling.

With regard to the suitability of the location of the housing site, Batas Pambansa (B.P.) 220[1] and its revised implementing rules by HULB require the following:

> … socialized housing projects shall be located within the suitable site for housing and outside potential hazard prone and protection areas (Rule 2, Section 4 [d]).

It also stipulates the physical characteristics of the housing site as follows:

> A potential site must have characteristics assuring healthful, safe and environmentally sound community life. It shall be stable enough to accommodate foundation load without excessive site works. Critical areas (e.g., areas subject to flooding, landslides, and stress) must be avoided (Rule 2, Section 5, A.3).

Aside from these requirements, P.D. 957 assumes that the developers of any subdivision and housing projects must follow the National Building Code (NBC) and zoning ordinances of the local government. Its Revised Implementing Rules and Regulations on easements of housing projects specifically requires that developers must comply with PHIVOLCS per Resolution No. 515, Series of 1992, on identified fault traces (Rule I, Section 1, B, 3.).

The PDRRMA states that the government must provide for the restoration and improvement of facilities, livelihood, and living conditions of disaster-affected communities including efforts to decrease disaster risk factors in accordance with the principles of "build back better" (BBB) for the PDR of disaster victims (Section 3, [aa]). This implies that the site of the resettlement must be, above all, conducive to the rehabilitation of disaster victims and safer from disaster risks compared with the previous disaster-prone residential areas of the victims. The UDHA,

[1]http://hlurb.gov.ph/wp-content/uploads/laws-issuances/mandates/bp_220.pdf.

which assists the PDRRMA in carrying out its long-term PDR goals, also provides that the relocation site where the socialized housing and resettlement projects are located must be near to areas where employment are accessible (Section 22).

5.4 The UDHA and the Role of the LGU in Site Selection

The UDHA has vast discretionary powers with regard to the relocation and housing of the urban poor living in danger areas as well as those who are victims of demolitions. Under the Local Government Code (LGC) of the Philippines, the local government, such as the municipal government of Rodriguez, exercises control, supervision, and administration over its jurisdiction.

The UDHA was enacted in 1992, one year after the enactment of the Local Government Code, which increased the authority and autonomy of the local government units (LGUs) to administer their jurisdictions. After the devolution trend, the UDHA was enacted with a view of empowering the LGUs to take care of the relocation and housing of their poor constituents. Thus, the UDHA mandated the LGUs to select the suitable location for resettlement and socialized housing in their territories. Under Section 7, the UDHA required all city and municipal government to conduct an inventory of all lands and to identify an appropriate site for socialized housing within their localities. The UDHA's LGU guidebook for resettlement and socialized housing specifies the following basic steps that LGUs must undertake to select the suitability of the relocation of the site (LGU Guidebook for Local Housing Project/Program, p. 13):

1. inventory and profile of available sites;
2. location and quality of site (accessibility to infrastructures and employment, quality of the site, land ownership, basic services, utilities, and cost;
3. location criteria of affected people and comparative analysis of possible sites (for resettlement sites); and
4. safe from both environment and man-made hazards.

However, this prerogative of the LGU to select the suitable location for relocation has been downplayed with a legal provision of the UDHA (Section 8), which requires that the LGUs, after the inventory, must coordinate with the National Housing Authority (NHA), the Housing and Land Use Regulatory Board (HLURB), the National Mapping Resource Information Authority (NMRIA), and the Land Management Bureau (LMB) before identifying the suitable site for relocation and housing. Moreover, this power of the LGUs is further diminished with the provision of the UDHA's LGU Guidebook, which requires that selection of site of the housing program must be based on the expressed preference of the locality's beneficiaries taking into consideration the site's access to transportation, provisions for basic services, and support facilities (p. 13). Despite this requirement of considering the expressed preference of the beneficiaries, it is ultimately the

NHA and its developers that determine the site for the relocation. Under UDHA, the NHA was tasked to undertake the identification, acquisition, and disposition of lands for socialized housing as well as to undertake relocation and resettlement of families with local government units. Because the government is enticing the private sector to invest in low-cost housing, the whims of its accredited developers with regard to site selection are often followed by the NHA.

This is in contradiction to the UDHA's provision that requires the expressed preference of the beneficiaries about the choice of relocation site. In the case of the SRHP relocation site, the municipal government is placed in a difficult situation where it is most likely blamed by beneficiaries and the public for any problems and defects in the relocation site because it is built within its territory. Although it has jurisdictions in the enforcement of criminal laws and estate management in the relocation area, it does not exercise administrative control and ownership of the site. The relocation is planned, constructed, and managed by the developers under the government-owned and controlled corporation–the NHA. All of the public biddings for the building of the houses, as well as the monitoring of the site and awarding of housing units to beneficiaries, are handled by this national agency.

The selection of the relocation site and the assessment of its suitability for disaster victims are also determined by the NHA. Without administrative power, the local government has no discretion with regard to the choice and determination of the site. The site location and suitability are therefore largely determined by what happened in the formal and informal deals between the NHA and the developers, of course, with the participation of some unscrupulous HULRB officials who approve the housing project. For example, there would be no cutting of corners—such as using substandard materials and construction defects in the relocation—if the contract and implementation were above board, free from interference of informal ties and agreements of unscrupulous intermediaries or between top housing officials and developers, and actively monitoreding by public regulators.

The discretion of the NHA officials, along with the influence of top government politicians, in dealing with private developers has resulted in negative consequences to the housing and relocation of disaster victims. This only indicates that semi-autonomous social fields are stronger than the official PDRRMA and other housing laws concerning the selection of the location and suitability of the resettlement area.

5.5 Hegemonic Powers of Private Developers in Site Selection

Under the implementing rules and regulations of the UDHA that governs section 18 (balanced housing development provision) of Republic Act (RA) 7279, private developers are required to set aside 20% of the total land (those selling lots only) and cost of their subdivision and condominium projects (those selling house and lots) for socialized housing for the poor. Developers can also comply with this legal provision by establishing a new settlement for the poor certified by the national and local governments, slum-upgrading certified by the NHA or LGU, participating in a Community Mortgage Program (CMP) where the developer finances the acquisition, development, and subdivision identified by the government, or entering into joint ventures with the LGU or government housing agency for socialized housing by providing 20% participation of the project area and cost (Section 3, a–e).[2] According to informants, if this provision of the law is seriously followed by developers and implemented by the government, there is no major problem with regard to the suitability of the relocation and low-cost housing for the poor and disaster victims. Most large developers have subdivision and condominium projects in Metro Manila and urban centers. They can accommodate many of the urban poor and homeless disaster victims in their housing projects and minimize out-city relocations that displace the beneficiaries from their urban jobs and informal trade. But to avoid mixing the rich and poor owners in one high-end subdivision and condominium project, which can affect the social stratification of owners, market value of the project, and project's marketing, developers, according to informants, prefer to follow the 20% percent rule by establishing low-cost housing projects in remote areas in coordination with the NHA and LGUs. With their strong connections and network ties with the NHA and other government housing agencies, developers can often get what they want to increase profit and circumvent this law of providing resettlement to the poor.

The power of private developers to ultimately choose the suitability of the site for socialized housing would not happen without the tacit consent of the NHA, which in turn could not happen without the government's undue benefits to private investors participating in public low-cost housing program. One unintended consequence of the government's campaign of enticing the private sector to participate in its socialized housing as well as the President's Benigno Aquino's Public–Private Partnership policy (PPP)[3] is the undue advantage or benefit given to the developers to the extent that they can undermine some of the major state laws and largely determine the quality of the government's relocation and low-cost housing projects. Despite the numerous laws to safeguard the government's housing interest and welfare of the poor beneficiaries of these housing projects, a few influential NHA-accredited developers with a strong connection with top public officials in the housing bureaucracy can reign as the leader or hegemon in the government's low-cost housing system.

---

[2]http://www.hudcc.gov.ph/sites/default/files/styles/large/public/document/Balanced%2020%20Percent.pdf.

[3]President Benigno Aquino III issued Executive Order No. 136 creating the public–private partnership (PPP) and amending Executive Order No. 8 (S. 2010), which reorganized and renamed the build–operate and transfer center to the Public–Private Partnership Center of the Philippines. Under the PPP, private businesses are given more power to participate in government projects with more incentives and financial and administrative assistance from the government.

The partnership between the government and private developers as advertised in government slogans is not actually a relationship between the dominant and the subordinate where the government remains in control of the entire low-cost housing process; rather it is "one among equals," and at times the government appears to be "dominated" by the private developers. In his 2015 State of the Nation Address (SONA), President Benigno Aquino III mentioned how the government pleaded—making it appear that the government is ready to accommodate the whims and caprices of private companies just to let them participate in the housing program—the private sector to invest in the government infrastructure under the administration's Public–Private Partnership (PPP) policy. In the low-cost housing sector, the national government has been giving private developers the best benefits—such tax breaks, marketing, financing, or technical assistance—through the NHA just to ensure that private companies earn enough to stay in the government's housing projects. With this full backing of the government, private developers appear to be hegemonic in their dealings with the NHA and other government housing agencies in determining on how the official housing laws and rules can be followed and bent to build relocation and housing projects that promise the highest return on their investment. The concessions given by the government through the NHA to private developers in joint-venture agreements have negative effects to the quality of housing projects extended to disaster victims such as the Typhoon Ketsana victims in SRHP.

5.5.1 *Joint Venture: The Contractual Relationship Between the NHA and the Developers*

The contractual relationship of parties in a joint venture, as the term implies, is one of partnership. As defined by the Philippine government's National Economic Development Authority (NEDA), a joint venture (JV) is "[a]n arrangement whereby a private sector entity or a group of private sector entities on one hand, and a Government Entity or a group of Government Entities in contrast, contribute money/capital, services, assets (including equipment, land, intellectual property or anything of value), or a combination of any or all of the foregoing to undertake an investment activity"[4] A JV is considered a species of partnership. When "two or more persons bind themselves to contribute money, property, or industry to a common fund, with the intention of dividing the profits among themselves," then a partnership is said to have been created. An ordinary partnership is organized for the general business venture and does not have a definite term of existence; whereas a joint venture is organized for a specific project or undertaking.[5]

[4]http://www.neda.gov.ph/wp-content/uploads/2014/03/2013-Revised-JV-Guidelines.pdf.

[5]https://attyarneldmateo.files.wordpress.com/2011/05/law_on_joint_ventures.doc.

Joint ventures are the primary method employed by the government to encourage private developers to join the public low-income housing production. Between 1987 and 2001, for instance, approximately 131 joint venture arrangements were formed under the government's joint-venture program (Ballesteros, 2012). This joint-venture arrangement offered by the government between the LGU, NHA and/or private developer can be of any of the following four types:

(a) A LGU owns the land, and the private developer assumes all aspects of development and housing construction with the financial requirement assumed by the developer and NHA.
(b) LGU owns the land whereas the financial requirement is shared with NHA. Development is contracted out to a private developer;
(c) A LGU owns the land and undertakes the development of site and housing construction, whereas the NHA provides the financial requirement.
(d) A private landowner owns the land, and a LGU undertakes development and shares part of the financial requirement with the NHA (Ballesteros, 2002, p. 30).

The most popular type of joint venture that relates to the relocation and housing of disaster victims, such that of Typhoon Ketsana victims, is either option A or option B with private developers having more control in the development and construction of the relocation and housing project.

The primary strength of a joint venture is that the NHA has been able to facilitate developments of its land as the private sector infuses investments into the project (Ballesteros, 2002, p. 26). However, in joint ventures, the government seems to lose some coercive and law-enforcement powers to private developers in exchange for their participation and investment in the housing sector. With this contractual relation, parties are considered partners or of "equal" standing although the government, through the NHA, remains the main source and guarantor of low-cost housing funds.

A critical weakness of the joint-venture approach which empowers private developers is that the NHA has no real influence on the determination of the results of a relocation project as well as the settlement location. "The only thing that binds the developer on the site selection is the Terms of Reference (TOR), which considers primarily environmental standards. Aside from environmental requirement, the TOR does not require assessment of the employment potential in the area or access to existing social services. As a result, private developers often offer cheap sites to the NHA that are unattractive to the formal housing market and are usually located far from the city centers" (Ballesteros & Egana, 2012, p. 8).

5.5.2 *Undue Benefits to Private Developers*

Enticing private developers to participate and invest in government low-cost housing does not happen without a price. Ballesteros (2002) argued that this benefit of the private sector infusing funds into the low-cost housing is outweighed by the cost and risk to government:

> First, NHA sometimes ends up doing the feasibility study, packaging and finalizing the project, which it also approves. Second, because NHA has a hand in reviewing the mortgage take-out before submitting to the NHMFC, it also shares in the accountability for accounts in default. Third, NHA also takes responsibility in cases when delays are caused by changes in policy of other shelter agencies. NHA has no control over the other agencies. Fourth, in the event that the partner is unable to collect from NHMFC, NHA also suffers from defaults from the bridge financing it has provided (Ballesteros, 2002, p. 26).
>
> Second, technical assistance pertains to NHA providing project packaging and engineering works (e.g., survey) and assisting in marketing. The assistance also includes an extension of financial accommodation to the partners of the program such as bridge financing, purchase commitment line with the NHMFC on buyer's financing and facilitation for developmental financing.
>
> Third, the NHA facilitates transactions related to other government agencies such as securing land conversion clearance, approval of permits and licenses and the utility companies. It is noted that 'technical assistance' has become a standard provision for all types of venture arrangement. Given the highly bureaucratic permit system, the delays in the release of payment from NHMFC and in finding the beneficiary that will qualify for mortgage financing under NHMFC, this technical assistance provided by the NHA significantly decreases the transaction costs of the private sector. These costs are often not reflected in the sharing of profits because such costs are mostly translated in terms of time. For instance, it takes an average of one year to obtain approval of permits and licenses.
>
> Clearly, the attractiveness of joint ventures mainly emanates from the provision of technical assistance for the bureaucratic problems (Ballesteros, 2002, pp. 25–26).

5.6 Laws on Accessibility

The rules refer to two kinds of accessibility: physical accessibility of the location and accessibility in terms of ownership of the housing project to the marginalized people such as disaster victims. Concerning the accessibility of the housing project, B.P. 220 and its revised 2009 implementing rules and regulation stipulate the following:

> The site must be served by a road that is readily accessible to public transportation lines. Said access road shall conform to the standards set herein of these Rules to accommodate

> expected demand caused by the development of the area. In no case shall a subdivision project be approved without a necessary access road/right-of-way constructed either by the developer or the Local Government Unit.[6]

P.D. No. 344, known as the Accessibility Law, even requires that the developer must provide provisions for disabled persons:

> [N]o license or permit for the construction, repair or renovation of public and private buildings for public use… shall be granted or issued unless the owner or operator thereof shall install and incorporate in such building, establishment, institution or public utility, such architectural facilities or structural features as shall reasonably enhance the mobility of disabled persons such as sidewalks, ramps, railings and the like.[7]

On site selection, the LGU's Guidebook for Local Housing Project/Program, prepared by the HUDCC, requires the following steps in choosing a suitable site for local housing:

1. inventory and profile of available sites;
2. location and quality of site (accessibility to infrastructures and employment, quality of the site, land ownership, basic services, utilities, and cost);
3. location criteria of affected people and comparative analysis of possible sites (for resettlement site,); and
4. safe from both environment and man-made hazards.[8]

In the selecting the site, the LGU's guidebook also requires that selection must be based on the beneficiaries' expressed preferences and that the site must have access to transportation and should have provisions for basic services and support facilities.[9]

Neither the PDRRMA nor the UDHA is clear on who exactly is responsible for allowing access to the SRHP relocation in the Municipality of Rodriguez: the NHA or the municipal housing agency—the HPDO? Or some other group? This ambiguity leaves more discretion for public officers to interpret the law, which sometimes leads to conflict at the expense of disaster victims. Based on interviews, it appeared in reality that the local governments—through the mayors of Metro Manila, some unscrupulous NHA officials, and their corrupt networks and the mayor of Rodriguez, Rizal, through the Housing and People's Development Office —could bring disaster victims to the SRHP. The local government of Quezon City, for instance, sent families who resided along danger zones such as the Tullahan and Culiat Creeks, the right-of-way along BIR roads, Barangay Tatalon, NIA road, and those affected by the Joint Venture Agreement (JVA) of NHA and AYALA or the

[6]See Rule II, Sect. 5, A, 4 at http://old.hlurb.gov.ph/uploads/laws-issuances/PD957IRR/IRRPD957.pdf.

[7]Section 1, The Law to Enhance Mobility of Disabled Persons. Available at http://www.architectureboard.ph/uploads/1331233623-Accessibility%20Law%20%28B.P.344%29.pdf.

[8]http://www.creba.ph/images/Housing%20Library/HUDCC%20LGU%20Guidebook%20for%20Local%20Housing.Pdf.

[9]Ibid, p. 11.

development of the North Triangle to the SRHP.[10] Aside from mayors, some unscrupulous officials of the NHA also sponsored people or professional squatters to the relocation site. According to the barangay administrator, there were instances of the illegal sale of housing units by some unscrupulous NHA officials in the relocation area, thus allowing professional squatters to enter the SRHP surreptitiously. The head of the Municipal Disaster Risk Management Council (MRRMC) recalled one instance that the Rodriguez police and the HPDO were able to intercept, based on a tip, five dump trucks, bearing NHA banners and loaded with professional squatters, heading toward the SRHP without the necessary entry documents and clearance. If disaster victims are residents of Rodriguez, Rizal, the mayor and the Housing and People's Development Office (HPDO) are authorized to allow them to be relocated to Southville Rodriguez after passing the pre-qualification process.

Aside from the SRHP, the local government of Rodriguez can provide its own disaster victims with socialized housing and livelihood programs. However, this all depends on the available budget, which in turn depends on the determination and creativity of the mayor in increasing funds beyond the 5% disaster fund for these purposes. During the incumbency of a former mayor, for instance, proceeds from the dump site were used to build a relocation site for local informal settlers, and the HPDO was able to provide livelihood programs for the beneficiaries (COA Report, 2002). According to the provision of the Local Government Code on the autonomy of local government units (LGUs), the HPDO and the mayor exercise discretion, monitoring, and supervision of the relocation area for attainment of their desired effects.

If the disaster victims come from areas outside Rodriguez and the municipality government has no available relocation area and funds, the situation becomes complex, and more unintended effects are expected because the local government no longer has total control of the disaster victims who are constituents of other mayors. Moreover, the Rodriguez government has no administrative control over the SRHP, being a relocation owned by the national government. This situaiton then requires coordination between the municipal government, other mayors, and the NHA concerning access to the relocation site. There are some requirements that need approval from the HPDO, such as the application for electricity or a water connection, even if the beneficiaries are non-residents.

The coordination problems are intensified when dealing with large-scale disasters, such as Typhoon Ketsana, with the new organizational set-up of the government disaster program under the PDRRMA. This coordination problem often causes an access delay for disaster victims to the relocation site and housing. Informants in SRHP 8A complained that they need to wait for months after the

[10]http://inclusivemanila.net/southville-8/.

typhoon before they can avail of the housing units.[11] While waiting for housing units, some evacuees stayed with their relatives, but many remained in a makeshift evacuation center in front of the barangay San Isidro government building. The local government of Rodriguez particularly encountered several coordination and jurisdictional problems in governing a relocation site managed by the government-owned NHA, albeit located within its jurisdiction such as the SRHP relocation site. Although the new PDRRMA explicitly intended that poor and homeless disaster victims must be given socialized housing and opportunities to lead a better life in a relocation area in accordance with the "build back better principle," the final outcome and implementation depend on the discretion and normative standards of the various actors and agencies in charge of socialized housing and disaster recovery.

5.7 Evaluating the Site Suitability and Accessibility of Southville Rodriguez

5.7.1 Site Suitability

The injunction of the LGU's Guidebook for Local Housing Project/Program on-site suitability provides that developers must do the following:

> [S]elect the site of your housing program based on the expressed preferences of your beneficiaries. Note, however, that the site should have access to transportation and should have provisions for basic services and support facilities (p. 13).

Although the law considers the expressed preferences of the beneficiaries as well as the site's access to transportation and basic services, it also provides that the priorities in the acquisition of the land for housing are the following type of lands, which are usually located in remote areas, unsuitable for housing, and of low market value:

1. idle government or private lands;
2. unregistered or abandoned idle lands; and
3. alienable lands in the public domain (p. 13).

This provision can be exploited by private developers to choose land for social housing that promises the highest return on investment. Despite the official law's insistence on consultation and participation of the LGU and beneficiaries in the site selection and development of the relocation and housing project and the site's

[11]During fieldwork in the SRHP last November 2012, I saw disaster victims in temporary evacuation shelters in front of San Isidro Barangay hall. I was informed by barangay officials that they had been staying there for months waiting for their housing application to be finally approved by the NHA. The NHA will issue them a stub denoting the block and lot number assigned to them before they can occupy the housing units in the SRHP.

accessibility to transportation, basic services, and support services, the private developer, who is in joint venture with the national government through the NHA, has the final decision on where to establish the relocation site. Owing to a lack of funds and technical capability to engage in low-cost housing projects, it is not the LGU but the National Housing Authority (NHA) and its private developers who have the power to determine the site of the resettlement. Ultimately, it is often the developer who decides which areas identified by the LGU will be used for relocation and housing projects, although the UDHA and the LGU Guidebook have given this privilege to the LGU and expressed preference of the beneficiaries.

With the policy shift of the NHA's role in land banking, the NHA identifies and selects resettlement sites based on the list provided by developers or LGUs. Site acceptability (on endorsement of community and local committees) is evaluated by the NHA based on a term of reference (TOR) that requires conformity with environmental standards and the subdivision standards based on BP220 (Ballesteros & Egana, 2012). Under the Completed Housing Resettlement Project or the developer-constructed approach, which has become a dominant mode in the government's relocation and housing projects, the NHA will simply choose from the developer's list of sites where the resettlement project will be constructed. Alternatively, the NHA-accredited developer(s) can offer (1) resettlement sites or (2) the sale of housing units in developed sites to the community association(s). The community selects the site and executes a formal agreement with the developer on the proper endorsement of the Local Interagency Committee (LIAC) and the NHA of the selected site. The NHA takes care of the financing scheme or loan finance, and the loan proceeds are paid to the developers on delivery of the housing units before the beneficiaries are relocated (Ballesteros & Egana, 2012). In this case, it is still the private developer who chooses the site for the relocation and housing. Because private developers are more motivated to maximize profit rather than the welfare of the beneficiaries, the land used for relocation is usually of low market value and unfit for the long PDR of disaster victims as demanded by the PDRRMA.

In the current housing and relocation set-up under the UDHA and NHA housing laws, the homeless disaster victims who need immediate shelters after a calamity have no real power to choose a relocation site that is suitable to their economic and cultural needs. In the case of Typhoon Ketsana victims of Rodriguez, Rizal, and some localities of Metro Manila, beneficiaries have no choice but to apply for housing units built in ready-made relocation sites constructed by NHA-accredited developers despite injunctions of the UDHA and its LGU Guidebook requiring the expressed beneficiary preference on the suitability of the relocation site and its proximity to employment and livelihood opportunities. Interviews with key informants in Phase 1A of Southville Rodriguez (SRHP) showed that some beneficiaries believed that their resettlement site is more hazardous than their former residential areas in danger zones.

5.7.2 From Danger Zones to Death Zones

A visit to SRHP Phase 8A, revealed more unforeseen negative effects to the actual PDR of Typhoon Ketsana victims. Key informants believed that the actual reality is far from what the official law provides. On paper, the law explicitly states that the relocation site for disaster victims must be safe, disaster-free, and suitable to their economic and cultural needs. However, in reality, the resettlement site is more risky, disaster-prone, remote, and far from employment zones. Many victims in the relocation even believed that they were moving from "danger zones" to "death zones":

> Napakahirap ang kalagayan namin dito. Inalis nga kami sa danger zone dahil malapit sa ilog ang aming mga bahay, subalit dinala naman kami sa pampang ng kamatayan. Mapanganib na lugar, marurupok na bahay, walang hanapbuhay, pagkain, malinis na tubig, at ospital. (Our situation here is very difficult. We were removed from the danger zone because our houses were built near the river, but they brought us to the banks of death: a dangerous site, sub-standard houses, no jobs, food, clean water, and hospital).

The official laws and rules expressly require that the relocation sites for disaster victims must be free from disaster risks. The intention of the PDRRMA and international conventions on PDR is to create communities resilient to disasters. This implies that the relocation must be safer than the danger zones where disaster victims resided before the typhoon. However, this seemed not to be the case. The Southville Rodriguez relocation site (SRHP) appeared to be more dangerous than the former residential areas of most of the Typhoon Ketsana victims. Table 2.1 shows that the key informants of this study lived in danger zones, residing in makeshift houses (*barung-barong*) near rivers and creeks and in low-lying areas, before they were relocated to the SRHP. The most common disaster risk they encountered in their residential districts before Typhoon Ketsana is flooding. All of them said that their makeshift homes were in low-lying areas located near a river. However, Southville Rodriguez is prone to various disaster risks aside from flooding (Table 5.1).

Ketsana informants believed that they were safer and more comfortable in their previous abode and location than in their current residential area. Because of the various disaster risks and problems in the relocation site, most Typhoon Ketsana victims in SRHP, especially in Phase 8A, believed that the government has transferred them from "danger zones" to "death zones" as one victim aptly describes:

> Niligtas nga kami ng gobyerno sa panganib, pero dinala naman kami sa hukay ng kamatayan: marurupok na bahay, mas mapanganib na lugar sa kalamidad, walang trabaho at hanapbuhay! (It's true that the government saved us from danger, but it brought us to death graves: weak houses, more dangerous place to disaster, no employment and livelihood.)

The United Nation's conventions and UDHA on adequate housing require that beneficiaries must participate in the process of selecting the relocation site. However, under the present set-up, the beneficiaries at Southville Rodriguez were

Table 5.1 Typhoon-Ketsana victims' places of residence and causes of flooding

Place of residence	Town/city	Frequency (N=27)	Reason for flooding
Madrigal compound, Sitio Tibag, *Barangay* Burgos	Rodriguez, Rizal	15	Near a river and a creek
Road 20, Nagtinig floodway; Purok 8, Barangay Arenda	Taytay, Rizal	5	Near a river and low-lying areas
Kasiglahan Village, Barangay San Jose	Rodriguez, Rizal	3	Near a river and low-lying areas
Sitio Masagoksok, Barangay San Rafael	Rodriguez, Rizal	2	Near a river
Barangay Tumana	Marikina City	2	Near a river

neither consulted nor allowed to participate in the choice of relocation site. The resettlement that was intended for the relocation of the urban poor was being used by the NHA for disaster victims. The power to identify and select the resettlement sites still belongs to the NHA. However, its choice is limited. The NHA can only choose from a list of relocation areas prepared by NHA-accredited developers or sites identified by the LGUs. Site acceptability (on endorsement of the community and local committees) is evaluated by the NHA based on a term of reference (TOR) that requires conformity with environmental standards and the subdivision standards based on BP220 (Ballesteros & Egana, 2012). However, based on interviews with local disaster officials of Rodriguez, Rizal, it is ultimately the developer who usually decides the choice of the relocation site in actual practice. Because of *palakasan* (patronage) or strong connections with top government officials, developers often get what they want from the NHA. They disregard the official law that requires developers to consider the endorsement of the community and environment clearance from the local government.[12] In the case of SRHP, even if the relocation area has not been endorsed by the beneficiaries and the MDRRMO because of its disaster hazards, the real-estate companies Baque and San Jose Builders—the developers of Southville Rodriguez—were able to obtain the

[12]According to Mercy Merilles, the spokesperson of the Montalban (Rodriguez) Relocatees Association (MRA), the national government, through the Department of Budget and Management (DBM), released PHP 10 Billion ($227) annually to fund government relocation programs and had, so far, disbursed PHP 27 Billion ($613.6) to government housing agencies such as the NHA. This fund constituted the Informal Settlers Fund, which, according to her, became the milking cow of corrupt NHA officials and source of business capital for low-cost housing firms. See: "What Safer Place? Urban Poor Relocatees Rebuffs Aquino", available at http://bulatlat.com/main/2014/07/16/what-safer-place-urban-poor-relocatees-rebuff-aquino/.

necessary permits and registrations for the project. With their strong connections in the government housing system, they often get what they want in order to maximize profit.[13]

The NHA employs different modalities for resettlement. In terms of method, relocation and housing projects can be classified as follows: (1) Completed Housing Projects (CHP) or developer-constructed projects; (2) Housing Material Loan or Incremental Housing Projects (IHP); and (3) NHA-LGU joint venture or the Resettlement Assistance Program (RAP-LGU). Between 2003 and 2010, the NHA executed the development of 88 resettlement projects nationwide, of which 70% (32 sites) of total projects were CHP- or developer-constructed. Under this method, the developer is assured of payment of all of the units constructed inside the relocation. However, its main critical weakness is the lack of real influence and control of the NHA on the settlement location. The only thing that binds the developer regarding site selection is the Terms of Reference (TOR), which primarily considers environmental standards but does not require an assessment of the employment potential or access to existing social services in the area. Thus, the NHA cannot directly intervene in the problem of employment and lack of socialfacilities and services caused by the developer's fault because this not included in the TOR. Moreover, the NHA has no choice if private developers offer cheap sites that are remote and unattractive to the formal housing market. The power to buy lands for settlement areas primarily belongs to private developers. The NHA only chooses from the developer's list of relocation sites. Of course, the NHA has the overall control of the housing and relocation system of the government. It can de-list or de-accredit problematic or abusive private developers. However, this is less likely to happen if the developers are *malakas* (powerful) or politically connected with top NHA and government housing officials.[14]

An investigation of the SRHP, where Typhoon Ketsana victims were relocated, revealed that the site is actually risky and prone to major disasters, which is contrary to the law's "build back better" principle (Table 5.2).

In an interview, the officers of Rodriguez's MDRRMO said that they were surprised why government regulators endorsed the construction of SRHP in Barangay San Isidro by accredited developers of the NHA despite the area's

---

[13]The owner of San Jose Builders, one of the developers of SRHP, for instance, is said to maintain powerful connections with Malacanang, the seat of power in the country. During the Estrada Administration, San Jose Builders became the developer of the huge Erap City or the Kasiglahan Village Relocation Area in Rodriguez, Rizal. As a sign of gratitude, the PCIJ reported that San Jose Builders gave Estrada the Boracay Mansion and renovated the home of Estrada's mistress in the Forbes Park Subdivision in Makati City. He seemed to continue to enjoy close connections with Malacanang because the President Aquino's executive secretary is said to be his brother-in-law. See: http://bulatlat.com/main/2014/07/16/what-safer-place-urban-poor-relocatees-rebuff-aquino/.

[14]San Jose Builders, as one of the developers of the SRHP Housing project, is said to have strong ties with top government officials including presidents. According to the Philippine Center for Investigative Journalism (PCIJ) report, it became the developer of a neighboring relocation site popularly known as Erap City because of its strong connection with the former Philippine President Joseph "Erap" Estrada: See Florentino-Hofilenia "Erap City" at http://pcij.org/stories/2000/erapcity.html.

Table 5.2 Disaster risks in Southville Rodriguez (SRHP)

Type of disaster risk	Description
Flood	Some parts especially those residing at the lower portions of Phase 1A
Earthquake	The entire Southville Rodriguez is located in between two West and East Fault Lines. Some fault lines also traverse the relocation site
Soil erosion	The land of Phase 1A is a reclaimed area form a former farmland. Thus, some parts are gradually eroding, thus causing cracks in housing units
Pollution	Southville Rodriguez is surrounded by various by mining, gravel, and sand companies resulting to dust on roads leading to the relocation site

geographical hazards. Being knowledgeable about the environmental conditions within its jurisdiction, the local government could have been more discreet in deciding where the relocation site must be situated. However, such is not the case with most low-cost housing projects of the government. It is usually the accredited private developers, with the backing of the NHA and top government officials, who exercise discretion on where to build the housing project for maximum profit. The present location of the SRHP is a former farmland surrounded by environmental risks. In fact, the entire relocation area is identified by the MDRRMO as being prone to landslides, soil erosion, earthquakes, and flooding. Some parts of the SRHP 8B and 8C, in particular, are flood-prone. Moreover, as early as 2004, an earthquake impact–reduction study predicted that the East and West Valley Fault Lines, which surround the present the SRHP area, could move at any time and create a devastating 6.5-to-7.9 earthquake that could heavily damage Metro Manila and surrounding provinces including Rizal and particularly the municipality of Rodriguez (MMEIRS, 2004). Why the NHA and its accredited developers were given environmental clearances, building permits, and registrations to establish the SRHP in 2009 is indeed puzzling to the MDRRMO officers, and one can only think of corruption as well as the Padrino (sponsorship) and *palakasan* (patronage) systems as the explanation for this discrepancy. Developers who have strong connections in Malacanang, HUDCC, HULB, and with NHA officials can develop and build relocation sites with the necessary permits, clearances, and registrations despite being declared by the local government as disaster-prone. Buying land in remote areas declared by the local government as disaster-prone is cheap and thus implies greater profit for developers.

5.8 Developer's Business Interest in Housing

Government slogans advertise that socialized housing is service and not a business such as *Ang pabahay ay serbisyo, hindi negosyo* (Housing is service, not busines) and *Gaganda ang Buhay sa Sariling Bahay (Life Improves in One's Home).* However, for the private developers in SRHP, building housing and relocation is primarily business, not a service, as illustrated by their action and involvement in resettlement and low-cost housing. The government's scheme of providing guaranteed take-outs or full payment to any housing unit built in the relocation area by developers, with or without household beneficiaries, has emboldened the latter to substantially decrease construction quality of the housing units and facilities in relocation in order to maximize profit:[15]

> The UHLP's formula enables developers to build and sell houses to eligible but not necessarily creditworthy borrowers without significant effect on their financial standing. "They earn sure income and do not have credit risk exposure," and at most they incur opportunity losses due to delays in government reimbursement through the takeout mechanism. In addition to the "moral hazard" of allowing developers to undertake mortgage origination, "the lack of credit risk, the ability to exploit information asymmetry in the production of houses and auxiliary infrastructure such as sewerage, piping, etc., and the guaranteed takeouts… create incentives for supplying substandard housing units and defective infrastructure" (Valte, 2013, p. 28).

Moreover, "[w]ith developer's objective for rapid turnover of capital, there is less incentive to find creditworthy buyers or to meet satisfactory development standards" (Ballesteros, 2011, pp. 2–3). Because low-cost housing is a business enterprise rather than a genuine exercise of social responsibility for disaster victims for private developers, all capitalist cost-saving strategies to decrease cost and maximize profit are therefore employed including the purchase of a remote and cheap land for relocation area, without regard to the suitability of the site to the safety and economic needs of the beneficiaries.

Government relocation areas in Metro Manila and nearby provinces are apparently of low market value and thus confer greater profit margin to the developers. Developers in low-cost housing projects tend to overprice their land for socialized housing projects because they are assured of payment from a "supportive" government. Some portions of Barangay San Isidro, where Southville Rodriguez was built, were declared disaster-prone areas and therefore unfit for housing projects according to informants of the Municipal Risk Reduction Management Office (MRRMO) of Rodriguez, Rizal, and thus would imply a very low market value. Yet the NHA allowed developers to build relocation sites and houses in these areas. Some key informants suggested bureaucratic corruption as the reason why the concerned government agencies allowed the developers to build the relocation despite the MRRMO's assessment. The Anakpawis Partylist group in the Philippine Congress, suspecting irregularities in low-cost housing, urged the United Nations to investigate and pressure the government to delist two corporate developers from accredited developers in the Housing and Land Use Regulatory Board (HLURB) for overvalued lands they sold to the NHA. These developers of SRHP were also involved in overpriced housing construction materials in the 2004 Towerville

[15] [R]equiring the financial system to take out or guarantee subprime loans implies that the government assumes 100% of the risk and that the resulting nonperforming loans translate to impaired assets and losses from the sale of these assets. The costs arising from the associated defaults of these loans amounted to more than PHP 1.0 billion in 2009 alone (Ballesteros, 2011, p. 6).

resettlement project (Ellao, 2011). Despite their tainted reputation, these powerful developers continued to win more contracts and became the contractors of SRHP in 2009 owing to their powerful social and political connections with top officials of the government's housing corporations. Because of *palakasan* and the *Padrino* system, these developers continue to sidestep the various official laws that regulate the government's housing projects to the ensure safety and adequacy of housing and relocation to the urban poor and disaster victims.

Developers usually prefer building projects in remote rural location because of low land valuation. Although the UDHA strictly requires developers to comply with the zoning ordinance of the local government in the choice of relocation site, the influence and illegal connections of large NHA-accredited developers with some unscrupulous officials of regulatory bodies of the government seems to disregard this requirement. The Baque Corporation, the developer of the SRHP Phase 8B, for instance, was served with an eviction notice by the City Mayor of Valenzuela for constructing a housing project without first complying with zoning and other construction laws (PCIJ, 2005). Building housing projects on remote and cheaper land could translate into huge profits for these developers. In the SRHP, developers sell a 40-m^2 lot with a 20-m^2 floor area of a bare housing unit to the NHA for PHP180,000.00 (around US$2,000) with or without an occupant as usually stipulated in project contracts. Any house and lot that developers build in the relocation site are automatically considered sold by the NHA. A COA report (2004) revealed that the objectives of government socialized housing were frustrated because many units remained unawarded to the intended beneficiaries. Under the present scheme, developers are assured of payment of their housing units from the government regardless whether or not they are awarded to beneficiaries. The burden of awarding the houses to the true beneficiaries, as well as processing and receiving the payments, is shouldered by the government.

Full payment by the government for every row house constructed in the relocation site enables the developer to continuously construct housing units without regard if there are already beneficiaries waiting to occupy them. A significant percentage of developers in the Philippines are small to medium scale firms who are after a fast rate of turnover of units and cannot wait for four years to recover their investments. This financial objective is even enhanced by government lending systems that require full cost recovery to ensure development loan repayment within a single project phase. The lack of subsidies and mass production of low-cost houses and lot packages that may be sold immediately by the developers to the government, resulted in downscaled and downgraded units that are not immediately livable" (Ramos, 2000, pp. 12–6) (Figs. 5.1 and 5.2).

Fig. 5.1 Front view of a brand new raw house in the SRHP Phase 8A relocation area. This house and lot is sold by the developer to the NHA for PHP 180,000.00 (around US$2,000) but the actual cost according to subcontractors and residents is probably between PHP 40,000 (around US$800) and PHP 60,000 (around US$1,200), given the low land valuation, construction defects, and poor-quality building materials. *Source* Author

Fig. 5.2 The interior of a brand new raw house in SRHP 8A. *Source* Author

5.9 Unintended Effects in the Relocation Site

Because the new PDRRMA did not provide specifics on the housing of disaster victims, thus allowing the old law to empower the NHA to manage relocation sites, more unforeseen effects can be observed in the relocation. On paper, the law explicitly states that the relocation site for disaster victims must be safe and disaster-free, but in reality the resettlement area is actually highly unsafe and disaster-prone. Some victims believed that they were safer in their previous place of residence than in the relocation area. As one relocatee of the neighboring relocation site of Kasiglahan Village Relocation site—popularly known as Erap City—which shares the same problem with the SRHP, aptly described the durability of the dwelling units in government relocation: "Konting bato, konting semento, sementeryo" (A handful of stones, a handful of cement, cemetery).[16] To increase profit, the private developer build very cheap houses, which are not only structurally weak and made of poor-quality construction materials, but are also built in a highly disaster-prone area.

5.9.1 Disaster-Prone Relocation Area

An in-depth investigation of the SRHP, where Typhoon Ketsana victims were relocated, revealed that the site is actually risky and prone to major disasters contrary to the law's "build back better" principle. As already mentioned, Barangays San Isidro and San Jose, where the SRHP is built, are listed by the Municipal Disaster Risk Reduction Management Office (MDRRMO) as two of the disaster-prone barangays in the Municipality of Rodriguez, Rizal, as shown in its geo-hazard map. These barangays are also flood-prone areas. The entire the SRHP relocation site used to be the catch basin of rain waters from the nearby mountains before they flow toward the Puray and Wawa Rivers. According to its head, the MDRRMO was able to secure a municipal ordinance to stop the developer from building more housing units in the SRHP Phase 8B due to its flooding problems. However, this was ignored by the developer because the local government had no jurisdiction in the management of the relocation area. It could only give a warning or advice on the disaster status of the site, but it could not directly intervene with regard to the construction policy on the site.

Flooding is not the only threat to the safety of disaster victims in the relocation area. The SRHP is built at the top of a minor fault line and surrounded by the two major fault lines of the West and the East Fault Lines. Building houses proximate to or residing on fault lines implies imminent danger to the life and property of the residents. The Philippines is located in the Pacific Ring of fire and is surrounded by fault lines. As already mentioned previously, a major earthquake mitigation study

---

[16]http://www.bulatlat.com/archive1/009erap_city.htm.

conducted in 2003–2004 under the auspices of the Japan International Cooperation (JICA), Metropolitan Development Authority (MMDA), and Philippine Institute of Volcanology and Seismology (PHIVOLCS) (MMEIRS, 2004) predicted that at any time a very destructive earthquake, with an estimated magnitude between 6.5 and 7.9, could hit Metro Manila and the neighboring provinces, including the SRHP area, owing to movement of the West Valley fault line. This fault line, according to PHILVOLCS, moves every 200 years; its last cycle has already elapses; and thus it is ripe for movement (Torregoza, 2011). With substandard construction materials, a weak soil foundation and construction defects, the SRHP Phase 8A housing units could easily collapse and result in deaths and loss of property in the area. This is indeed the most serious unintended consequence of the housing and relocation program not being contemplated by the PDRRMA and its affiliate laws (Fig. 5.3).

5.9.2 *Other Unintended Effects*

Because the relocation area is remote and far from urban employment, some beneficiaries have sold their rights to the housing units. Under the law, the beneficiary cannot sell his or her housing units. This will automatically disqualify him or her for future housing programs of the government. However, according to informants some units that were not occupied by the beneficiaries—because they returned to their original residence because of employment opportunities or because of extreme poverty were sold—were sold for PHP 40,000–80,000 depending on the condition of the house. This act is considered the sale of rights only because the houses were still not fully paid for by the beneficiary. However, this defeats the purpose of the housing program. The new occupants or buyers of these units were not really poor and disaster victims. Some of these were professional squatters or members of housing syndicates who buy and sell or lease the housing units for profit. One buyer from Rodriguez, Rizal, for instance, revealed that he bought one housing unit in the relocation site for PHP 80,000.00. He is neither a disaster victim nor really poor because he works as a regular employee in a private company.

Those who keep their homes in the relocation area and continue to work in Metro Manila usually return to the relocation site only on weekends. During weekdays, they work as informal workers or street vendors in urban centers. To minimize transportation cost, many opt to sleep in public places to avoid paying room rent according to informants. Mang Renato, for instance, works as a carpenter in construction in Quezon City from Monday to Saturday and returns to the SRHP on Saturday evening and Sunday. He is lucky because he can sleep at the construction site. However, Aling Mila, who sells fruit on the sidewalk of Cubao area, slept in public places at night (Fig. 5.4).

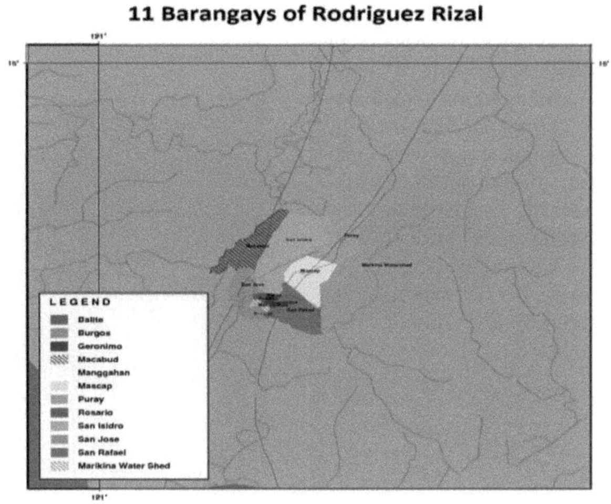

Fig. 5.3 Map showing the five barangays where Southville and Erap City are located San Isidro, San Jose, Macabud, Puray, and Mascap as well as the two major East and West Valley Fault Lines (*red lines*) surrounding them

5.9.3 Access to Southville Rodriguez

5.9.3.1 Kanya-Kanya and Access to the Relocation

When it comes to disaster management, the local governments, especially barangays and municipalities, have a crucial role to play because they are the front liners who directly implement the law. In fact, the Barangay Disaster Risk Management Committee (BDRRMC) and Municipal Disaster Risk Reduction Management Office (MDRRMO) personnel do the actual rescuing of disaster victims, i.e., assisting them with their immediate needs, placing them in evacuation sites, and processing their applications for relocation and housing. The UDHA, which implements the relocation provision of the PDRRMA, provides that there must be coordination between various government agencies for resettlement of the urban poor. However, this seemed not to be the case in actual implementation of the law. Each government agency has its own "turf" and tends not to share resources with others. The Filipino value of *kanya-kanya* (following one's own group)often occurs in governmental coordination. In the case of the SRHP relocation, municipal disaster officials are more familiar with the problems and needs of their constituents and disaster victims than the officials of a national agency such as the NHA. And yet the NHA, which holds administrative control of the relocation area, does not often listen to the suggestions and recommendation of local-disaster officials. Interviews with the Barangay administrator of San Isidro and the head of the HPDO showed that the NHA decides things for the SRHP without consultation and proper coordination with the local government or *kanya-kanya* in doing transactions that can affect both the national and local governance.

For one, the NHA does not provide the local officials with the proper information or demographic data of the people relocated in the SRHP, especially

Fig. 5.4 A long shot of the rock riprap at the back of the SRHP. Some parts of this relocation are reclaimed areas of a farmland. Soil instability is a major reason why some parts of the relocation are gradually sinking, thus causing cracks in the houses. The area shown here is usually flooded during strong storms. *Source* Author

beneficiaries who do not come from the municipality. According to the HPDO Head, mayors from Metro Manila who bring disaster victims into the relocation area also practice *kanya-kanya* in sending relocatees to the SRHP. Instead of coordinating with the municipal mayor or the HPDO office of Rodriguez, Rizal, for the entry of non-resident relocatees, they go directly to the NHA to negotiate and, at times, directly to the Office of the President in Malacangan Palace and send their relocatees to Southville Rodriguez without providing HPDO officials with the profile and basic information of the new beneficiaries. The NHA, in turn, despite knowing these negotiations, does not also immediately inform and update the HPDO with regard to new migrants entering the municipality. Moreover, the NHA does not regularly update the HPDO concerning new construction or administrative changes in the relocation. Up to the present, the NHA has consistently resisted, according to the HPDO, in providing a master list or profile of all the beneficiaries or relocatees inside the SRHP for monitoring purposes in order to prevent multiple ownership of housing units and entry of professional squatters and illegal groups in the relocation.

The absence of a master list of all residents in the SRHP has left the LGU or HPDO clueless with regard to the entry of illegal or non-resident relocatees in the relocation area. The Barangay Disaster Risk Reduction Management Council (BDRRMC) officials in San Isidro, who directly supervise the resettlement in terms of disaster management, socio-economic development, and peace and order, are also groping in the dark on the demographic profile of non-Rodriguez disaster victims who migrated to the relocation site. Thus, according to the barangay administrator of San Isidro, a census was recently commissioned in order to obtain these data. Local officials complained of the lack of cooperation of the NHA about the sharing of information and administration of the SRHP which technically belongs to their local jurisdiction.

Although the PDRRMA seemingly empowers local governments in disaster management, local officials appear powerless in the case of the post-disaster recovery of disaster victims at the SRHP. Even with mayors negotiating with the NHA and with top government officials, the BDRRMC and HPDO could not control the influx of disaster migrants in the barangay and in the municipality. According to barangay officials, the infrastructures of San Isidro could no longer accommodate the increasing population in the SRHP. It lacked funds to develop projects and to hire more barangay personnel, such as *tanod* (police) or health workers, to attend to the needs of the local population especially with regard to peace and order and health services. According to them, the continuous influx of migrants in San Isidro has also affected the services and infrastructure of the barangay. The public utility vehicles in the relocation area, like tricycles and passenger jeepneys, are still illegally operated without franchises from LGUs and the Land Transportation and Franchising Regulatory Board (LTFRB). Major public works in the area near the vicinity of the relocation remained unattended because of the lack of funds. The main entrance road to the relocation site, for instance, remains rugged and uncemented. For San Isidro barangay administrators, the local government must be consulted first before additional relocatees can be allowed into the SRHP Rodriguez (Fig. 5.5).

However, this restriction seems not to be the case. One reason why mayors and relocatees adopt *kanya-kanya,* i.e., bypassing the entry requirements of the local barangay, is political in nature. Non-residents of Rodriguez who relocated in the SRHP are required by barangay officials to register as voters of the municipality. For barangay officials, registration makes these people legitimate residents of the municipality and thus can avail some of the benefits accrued to local residents. However, for some relocatees, registering as voters before entering the relocation site is a strategy of the mayor and other officials to obtain more votes during local elections. Barangay San Isidro is a vote-rich area in Rodriguez, Rizal, because of the ongoing relocation and migration. A mayoralty candidate who can win in San Isidro has a greater chance of being elected as mayor. In fact, the incumbent mayor is the former barangay captain of San Isidro.

Fig. 5.5 The rough road at the main entrance of the SRHP. *Source* Author

Mayors of Metro Manila or nearby municipalities who bring relocatees to the SRHP want to bypass this registration process to avoid losing votes from their constituents; thus, the *kalakaran* (practice) of *kanya-kanyang diskarte* (individual strategy) of bringing relocatees to the SRHP while avoiding coordination as much as possible with the local government or municipal and barangay officials. For migrants, retaining their residence status is beneficial to them. According to informants, migrants prefer to retain their residence status because they do not want to lose their benefits from their original local government. Relocatees from Marikina City, for instance, do not want to lose educational benefits for their children or their affiliation with the mayor's local party because of monetary or employment benefits.

5.9.3.2 *Palakasan* and *Pahirapan* in the Pre-qualification Process

The accessibility of the housing and relocation program is influenced by social and political connections in the case of the SRHP. In the Filipino value system, *palakasan* and *pahirapan* (difficult process) are related but belong to opposite poles. People who are *malakas*, i.e., those who have the means and strong connections with government officials, do not necessarily experience *hirap* of getting what they want. However people who are *mahina* (politically weak), i.e., those with no connections with powerful people, experience *hirap* (difficulty), and the process of getting what they want then becomes *pahirapan* or "of extreme difficulty." The classic definition of politics by H. Lasswell as "who get what and when" can be applied here. Being poor, homeless, and victims of disaster is not a guarantee to become beneficiaries of the socialized housing of the government as stipulated by law. Disaster victims, such as those of Typhoon Ketsana, are required to undergo a tedious and costly screening or pre-qualification process performed by various local and national agencies before being awarded housing units in the SRHP. The final outcome would depend on the appreciation of the application and the degree of political connection of the applicant with government officials. In this sense, the housing program and relocation are not accessible to all poor homeless victims because *palakasan* influences the screening and application system.

The screening process starts when disaster victims file their application in the barangay. As stipulated by the 2010 PDDRMA, the Barangay Risk Reduction Management Council (BRRMC) receives the applications and conducts an initial investigation of the applicants. In the case of Typhoon Ketsana victims of Rodriguez, Rizal, the applicants must submit their applications to the BRRMC officials in Barangay San Isidro where the SRHP is located. In choosing the beneficiaries of the relocation and housing, the BRRMC uses the following criteria under the LGU's guidebook:

1. Poor families qualified for relocation and resettlement and assistance under the Urban Development Housing Act of 1992 (RA 7279).

2. Underprivileged and homeless citizens whose average or combined family income fall within the poverty threshold.
3. Those who do not own any real property and have not been beneficiaries of any government housing program except on leasehold or rental arrangement. First Priority—Target project beneficiaries Second priority—Open Market.[17]

To avail of the NHA housing, the applicant must present: (1) barangay clearance, (2) a marriage contract, and (3) valid identification.

After this initial screening, the Barangay Risk Reduction Management Council (BRRMC) makes a master list of qualified applicants and submits it, together with the application and supporting documents, to the Housing and People's Development Office (HPDO) in the municipal hall. The HPDO screens again the applicants whether they can qualify as beneficiaries in the socialized housing of the UDHA and creates a final master list to be submitted to the NHA for final approval. Interviews with HPDO and BRRMC officials revealed that the pre-qualification process is required by the UDHA for socialized housing applicants to prevent professional squatters or members of housing syndicates disguising as homeless urban poor or disaster victims to exploit the government's urban poor housing program.[18] Under Section 16 of the UDHA, only the following are eligible to be socialized housing–program beneficiaries of the government in relocation area as follows:

(a) must be a Filipino citizen;
(b) must be an underprivileged and homeless citizen,
(c) must not own any real property whether in the urban or rural areas; and
(d) must not be a professional squatter or a member of squatting syndicates.

Although the pre-qualification process is intended to weed out professional squatters, the complexity and rigidity of the application system sometimes create unintended effects that tend to frustrate the objectives of the new disaster law. For one, this costly and time-consuming process became an additional burden to their current application sacrifices. The victims do not only experience standing in long queues in the municipal hall, but also in the local NHA office in Kasiglahan Village usually under the strong heat of the sun. Two poor and homeless Ketsana victims recalled their difficult experiences with their applications:

---

[17]LGU's Guidebook, p. 14.

[18]"'Professional squatters' refers to individuals or groups who occupy lands without the express consent of the landowner and who have sufficient income for legitimate housing. The term shall also apply to persons who have previously been awarded home lots or housing units by the Government but who sold, leased or transferred the same to settle illegally in the same place or in another urban area, and non-bona fide occupants and intruders of lands reserved for socialized housing. The term shall not apply to individuals or groups who simply rent land and housing from professional squatters or squatting syndicates" (UDHA, Sect. 3 [m]).

Aling Berta:

Ilang araw na pagpipila sa munisipyo. Nakahihilo po sa pagkuha ng mga requirements. Lakad-takbo po ang ginawa naming para makauna. Mahirap pong makakain.... [A few days of queuing in the municipal hall. Getting the requirements can cause dizziness. We walked and ran in order to be first. It was difficult to eat.]

Aling Nena:

Madalas pabalikin sa NHA at pumila ng napakahaba sa init ng sikat ng araw at walang pamasahe o pangmeryenda man lang. [We often went to the NHA back and forth and followed long lines under the heat of the sun and without fare and snacks.]

Moreover, the frequent follow-up of their applications also drained their limited resources of the poor applicants. For Typhoon Ketsana victims in SRHP, the application and pre-qualification processes have intensified the suffering of the disaster victims. After having lost everything after the disaster, they must bear the inconvenience of long queues and hunger to obtain and submit the requirements.

5.9.3.3 The *Padrino* System, *Palakasan*, and Access to Housing

The most painful part of the pre-qualification process is the uncertainty of approval. Some of the informants complained of *palakasan, kakilala*, and the *Padrino* system involved in the awarding of housing units in the relocation site. Normally, the processing of an application takes 1–3 months before one can occupy a house in the SRHP relocation area. However, if one has a sponsor (*Padrino*), usually a politician, the waiting period is said to be shortened. Even those who are not qualified under the law can avail of the housing units in the relocation site if they have political backers. Two of the informants, for instance, attributed the fast approval of their applications to the sponsorship of a local municipal official whom they campaigned during the election. Thus, poor disaster victim who has no *kakilala* or *Padrino* in the pre-qualification process has a high probability of disapproval of their application and must go through the difficult process of securing the necessary documents for their application and a longer waiting period for their approval.[19]

The most alarming unintended effect in the pre-qualification process, according to some informants, is disapproval despite complying with all of the requirements and knowing that one is truly qualified under the law. Because local disaster and NHA officers have the discretion to approve or reject applications, *palakasan* often influences the selection process. Those who are *malakas* or have a political sponsor in the local government or NHA can easily avail of housing unit, but those who are weak with no connection must undergo the long route or *pahirapan* just to get a unit in the relocation site. The *Padrino* system sometimes influences the awarding

[19]Some informants revealed that they had difficulty securing the documents because the flood had washed away all their belongings. Moreover, it is more expensive for them to go to different government agencies just to obtain a copy of certificates for their application.

of housing units. The beneficiary must have a sponsor or *kakilala* (acquaintance) in the barangay, local DSWD, or municipal government to guarantee swift approval of their housing application. This was confirmed by three Ketsana relocatees in SRHP 8A, who were not actually poor and homeless and have real property, according to an interview, but were able to get housing units in the relocation because of their political sponsors. Local politicians—such as mayors, counselors, or barangay officials—often sponsor some people, especially their political supporters, to secure housing units in the SRHP despite being disqualified under the UDHA law. Even one barangay councilor who has a connection with the local municipal mayor was able to secure a housing unit in the SRHP.

The law specifically requires the LGUs to identify the legitimate disaster victims or homeless as beneficiaries of the government's relocation and housing projects. However, the *kalakaran* (custom) to secure housing units in the SRHP, according to the HPDO Chairperson, before he took over his post was that application forms were reproduced, photocopied, and distributed to the public and not exclusively for disaster victims. This means that anyone who has an application form and connection or *kakilala* (acquaintance) or *Padrino* (sponsor) or who has the money to bribe local officials can avail of the socialized housing intended for the urban poor and disaster victims. The law, which says that only the poor and homeless with no real property can avail of the relocation and housing, is seldom followed in practice. The informal rules of *palakasan, the Padrino* system, and *lagay* (bribe) dominate the system with regard to the accessibility of the resettlement and housing in government relocation sites such as the SRHP. In an interview, the HPDO Chairperson acknowledged that he saw names of relatives of some municipal officials, who are neither poor nor disaster victims, awarded with housing units in the relocation during the previous administration. He continued to see applicants who are not victims of disasters and who owned real property applying in his office for new houses in the SRHP. In the absence of a database or profile of the applicants and disaster victims due to the lack of computerization of government records, the HPDO usually faced with the difficulty of verifying the authenticity of entries in the application form, particularly with regard to the applicants' social class and ownership of real property. The Philippines lags behind in the computerization of government records and e-government compared to its Asian neighbors. An online database showing real property owners in the Philippines does not exist in the NHA records.

The official rules found in the UDHA and HUDCC's LGU guidebook require that the local government must actively identify and organize the beneficiaries of the government's housing program. The UDHA requires LGUs to "identify and register all beneficiaries within their respective localities," whereas the LGU guidebook requires the local government must first "conduct census of potential beneficiaries of the projects" and then organize them in preparation for their access to the resettlement and housing project (LGU Guidebook, p. 916). With these provisions, the application for relocation and housing must therefore be given only to the community of genuine beneficiaries and not to outsiders and disqualified individuals or groups. Under the PDRRMA, the beneficiaries must be, first of all, disaster victims who are underprivileged, homeless, and not professional squatters and owners of real property in the country (UHDA Section 16).

However, this seems not to be the case in actual practice. According to informants, the role of the LGU in identifying beneficiaries particularly in Rodriguez is not proactive as required by the official law but passive. In other words, the HPDO does not go to the beneficiaries; the beneficiaries go to the HPDO. The potential beneficiaries must approach the HPDO and submit applications for approval. In fact, the HPDO distributes application forms and waits for the disaster victims or target beneficiaries to apply. This process, according to informants, results in abuse of the relocation and housing system intended for the homeless urban poor and disaster victims by illegal groups and housing syndicates. According to the current HPDO head, some enterprising individuals who are familiar with the system and have connections with national and municipal officials sometimes bribe NHA or/and HPDO officials to obtain housing units in the SRHP. They pay a *lagay* (bribe) for each unit and sell it interested buyers who are not necessarily poor or disaster victims.

Housing syndicates are not gun-toting or organized crime groups but rather enterprising people who know the loopholes in the housing system, have strong connections with municipal and NHA officials, and engage in buy-and-sell or lease of government housing units in the relocation for profit. These syndicates operate at both the local and national level. Those at the national level are syndicates who acquire their housing units directly from the NHA, whereas the local syndicates obtain their housing units from unscrupulous local housing officials. Each housing unit has a corresponding bribe, according to informants, but the syndicate still earns huge profits when they sell the units to interested buyers at a higher price. According to informants, these housing syndicates usually have political connections with some officials in the NHA and even in Malacanang. Because bonafide beneficiaries in the SRHP do not possess legal documents as proof of ownership, only entry stubs, and because the NHA is non-transparent concerning the master list of the real beneficiaries or occupants of the relocation, there are instances where the units occupied by residents are actually under the names of other people.

5.9.3.4 Social Access and Political Patronage

The reliance of the PDRRMA on the pre-existing legal structure of the UDHA and the government's socialized-housing program has not only led to the inheritance of various unintended consequences not directly related to post-disaster recovery, but it has led to political patronage becoming embedded in the Philippine socialized-housing system. In a relocation site owned by the national government relocation site through the NHA, the LGU, according to the HPDO, is entitled to 10% of the total housing units. Thus, if Southville Rodriguez is expected to reach 60,000 housing units, the municipal government through the HPDO is entitled to 10% of 60,000 or 6000 housing units for their resident disaster victims or relocatees.

Despite the official law specifying that the LGU, where the relocation and housing project is located, is responsible for selecting the beneficiaries, there is an informal rule or internal arrangement within the provincial government that 10% of the housing units will be divided into three equal shares each to the municipal mayor, the congressman of the district, and the provincial governor. This means that access to the relocation and housing units pass through the approval of these government officials. One way or another, the beneficiary must be socially or politically connected to these officials, or to their subordinate allied officials such as barangay officials, to gain easy access to the relocation site and housing benefits. The disaster victims must therefore have a political sponsor or *Padrino* who is politically connected with these officials to expedite his or her application for housing in the SRHP. One *kagawad*, or barangay councilor, was able to relocate a group of Typhoon Ketsana victims from Barangay Manggahan, Rodriguez, Rizal, to the SRHP because of his close ties to the former municipal mayor. During elections, the barangay captains and officials such as councilors play an important role in delivering votes for mayors, governors, and congressmen. Thus, barangay politics is also important with regard to the endorsement of housing applications to the HPDO. The HPDO officials themselves, who are appointed by the incumbent mayor, also consider political affiliation aside from the informal rule of kinship ties (*kamag-anak*) or bribery (*padulas/lagay*) in approving housing applications for the relocation site. Disaster victims and applicants who are not political supporters or members of the in-group (*sakop*) of the incumbent local officials and who are not *kakilala* (acquintances/friends) of the barangay officials can encounter serious difficulties in availing the housing benefits of the government as envisioned by the PDR provisions of the PDRRMA. The strong informal and cultural normative sytems of different public and private stakeholders can sideline the legal norms of the UDHA and PDRRMA on relocation and housing for the poor and disaster victims.

5.9.4 Summary

This chapter has shown that despite the official laws' injunction to allow beneficiaries to determine the suitability of the site for their relocation and housing project, the power to choose the relocation site ultimately belongs to the NHA-accredited private developers, especially in the case of Typhoon Ketsana victims of Southville Rodriguez. The inadequate legal provisions of the PDRRMA and the government's joint-venture and PPP strategies to entice the private sector to invest in low-cost housing results in the adoption of the pre-existing arrangement regarding relocation and housing under the UDHA and undue benefits and hegemonic powers to private developers, especially in determining the suitability of the relocation in the developer-constructed projects such as the SRHP. Medium-sized private developers, assisted by the NHA in their technical housing needs and motivated more to maximize profit rather than providing a suitable site for the welfare of the beneficiaries, usually choose sites that are remote, disaster-prone, and of low market value, which is contrary to what the official laws of the PDRRMA and UDHA require. The SRHP is situated in a disaster-prone area of Barangay San Isidro and remote to urban-employment zones. This chapter has also shown that the criteria for physical and social access to the relocation as demanded by official international conventions and the

UDHA are largely unmet. The relocation is physically inaccessible due to its remote location and poor transportation system. It is also highly disaster-prone and remote from urban employment and livelihood activities of the beneficiaries. Social access is also not open to all of the disaster victims severely affected by Typhoon Ketsana. The informal rules and the cultural norms of *palakasan, lagay, padulas, Padrino system, kakilala,* and other social normative systems dominated the relocation system in the SRHP and sidelined the legal norms of the UDHA and PDRRMA that protect the poor and disaster victims. Those who had strong political ties with some top national and local politicians and public officers officials were more likely to gain access to housing and relocation area than those who were poor and truly affected by Typhoon Ketsana.

References

Ballesteros, M. (2002). *Rethinking Institutional Reforms in the Philippine Housing Sector.* Discussion Paper Series No. 2002–16. November 2002. Philippine Institute for Development Studies (PIDS). http://dirp3.pids.gov.ph/ris/books/pidsbk05-ppshousing.pdf

Ballesteros, M. (2011). Fiscal costs of subsidies for socialized housing programs: an update. *PIDS Policy Notes,* 2011–14.

Ballesteros, M., & Egana, J. (2012). *Efficiency and Effectiveness Review of the National Housing Authority (NHA) Resettlement Program.* http://www.dbm.gov.ph/62C97C27-20B5-4CB0-9BD2-8286EDF12907/FinalDownload/DownloadId-802ACD5E5502BED20A5F66A855713/62C97C27-20B5-4CB0-9BD2-8286EDF12907/wp-content/uploads/DBM%20Publications/FPB/ZBB-2012/e.pdf. Accessed March 4, 2015.

COA Report. (2002). Socialized Housing Program, Municipality of Rodriguez, Province of Rizal (Management Services, Report No. 2002–06, Sectoral Audit). http://www.coa.gov.Ph/GWSPA/GWSPA.htm

COA Report. (2004). *Sectoral Performance Audit Report on the Housing Program of the Government.* CYs 2001–2004. http://www.coa.gov.ph/index.php/sectoral-performance-audit/category/244-housing-program-of-the-government?download=701:national-housing-authority-ms-report-no-2004-02a

Ellao, J. (2011). *Urban poor urges United Nations to investigate rights violation*, published Oct. 25, 2011. http://bulatlat.com/main/2010/10/25/urban-poo-urges-united-nations-to-investigate-rights-violations/. Accessed 20 Jan 2013.

MMEIRS. (2004). Earthquake Impact Reduction Study for Metro Manila, Republic of the Philippines, Final Report, (Vols. 1–6). Tokyo: Japan International Cooperation Agency. http://www.ndrmc.gov.ph/Attachments/020_E_SUMMARY_01_PDF

PCIJ. (2005). City of Valenzuela, Office of the Mayor, Executive Order No. 2005-022, Series 2005. Immediate Order for the Ejectment of the Occupants of a Socialized Housing Project Owned by Morningville Homeowners Association Developed by Baque Corporation. http://pcij.org/stories/2005/Gatchalian-eviction-oder.pdf

Ramos, G. (2000). *The Urban Development and Housing Act (UDHA) of 1992, A Philippine Housing Framework.* http://www.lth.se/fileadmin/hdm/alumni/papers/ad2000/ad2000-12.pdf

Torregoza, H.L. (2011). *Marikina Valley Fault Line _ripe for movement' PHIVOLCS.* Manila Bulletin Publishing, p.1, March 16, 2011. http://www.mb.com.ph/node/309766/marikina-valley-fault-line-ripemovement-phivolc

Valte, M. (2013). *A House for Every Filipino: A Pipe Dream in the Land of Promises?* http://www.ombudsman.gov.ph/UNDP4/wp-content/uploads/…/Chap6.pdf

Chapter 6
Security of Tenure, Affordability, and Habitability

Abstract The right to adequate housing includes the right the security of tenure to one's housing unit and the right against eviction. It also requires that the low-cost housing for disaster victims must be affordable and commensurate with the income of the urban poor. Above all, the housing units must be habitable and resilient to disasters. This chapter examined how these rights are protected and implemented in the relocation and housing program of Typhoon Ketsana victims in Southville Rodriguez Housing Project (SRHP) in Rodriguez, Rizal, in accordance with the 2010 Philippine Disaster Risk Management Act (PDRRMA) and related laws. It also examined how the informal connections and rules between accredited developers and the National Housing Authority (NHA) and other housing agencies influenced the implementation of the official laws on post-disaster relocation and housing. It explained why the official laws, which were intended to build affordable and habitable homes for disaster victims with occupants enjoying the right to security of tenure, turned out to be unaffordable and unfit for human occupation due to their weak construction materials, defects, and soil erosion.

Keywords Adequate housing · Security of tenure · Affordability · Habitability · Post-disaster housing · Relocation · Eviction

6.1 Security of Tenure

The intention of the law in creating "resilient" communities implies that disaster victims must not only be given permanent homes, even better than their previous ones, but also legal ownership and security of tenure against evictions or harassments. The Urban Development and Housing Act of 1992 (UDHA), under Sect. 3 (o), defines security of tenure in socialized housing as "the degree of protection afforded to qualified Program beneficiaries against infringement or unjust, unreasonable and arbitrary eviction or disposition by virtue of the right of ownership, lease agreement, usufruct and other contractual arrangements." Under the Philippine civil law, the conclusive and indisputable proof of security of tenure is

V.O. Ballano, *Law, Normative Pluralism, and Post-Disaster Recovery*,
DOI 10.1007/978-981-10-5074-9_6

the Torrens title. If the house and lot are bought on installment, the contract to sell executed between the buyer and seller becomes a reliable proof of ownership and security against eviction while the property is still in mortgage and has not been fully paid for by the buyer.

Although the PDRRMA does not explicitly state the right of disaster victims to their homes in relocation areas, it nevertheless affirms the right of occupants against eviction. It implicitly recognizes the right of security of tenure of relocatees to their housing units by stating that it upholds the constitutional right to property in socialized housing (Sect. 1, [b]) and "incorporates the internationally accepted principles of disaster risk management in the creation and implementation of national, regional and local sustainable development and poverty reduction strategies" (Sect. 1, [c]). This implies that the PDRRMA does not only recognize all pertinent laws on titling and security of tenure but also adopts the United Nations' (U.N.) understanding of security of tenure. Under the U.N. conventions, security of tenure has a broader scope. It not does only include the proof of ownership but also the right against eviction. According to international human rights law, the U.N. also understands the right to adequate housing as the right to protection against arbitrary or unlawful interference with privacy, family, home, and to the legal security of tenure (UN Guidelines on Housing, Annex 1, p. 5).

6.1.1 Security of Tenure in the Southville Rodriguez Relocation Area

Disaster victims who have lost almost everything are entitled to feel a sense of security that after paying 30 long years of amortizations they can finally own a home. This is not what the Ketsana victims in the SRHP Phase 8A, believed with regard to the legal security of their homes. Disaster relocatees in the relocation area complained to local officials that the NHA only gave them an entry pass, a small piece of paper bearing the name of the beneficiary and the address of the unit, as proof of ownership to their housing units. They have no copies of title to the land and house, just the orange-and-green entry pass issued to them by the National Housing Authority (NHA). In an investigation by Demolitionwatch in the SRHP Phase 8B, it was discovered that residents did not have a proper contract or agreement with the developers regarding the ownership of their housing unit. Home owners said that they had signed the agreement but did not fully understand the content of the contract with the NHA. In fact, when asked how much is the contract price of their housing unit, their answers varied and estimates ranged from "PHP 100,000 to 200,000" (US$2000–4000) indicating that they were not fully aware of the contract price of the housing units awarded to them by the NHA.

Although Typhoon Ketsana victims were required to sign documents in the local NHA office in Rodriguez, Rizal, before being given entry stubs, no duplicates or photocopies of the original contract to sell or other documents as proof of ownership of their housing units were given to them. Because of their lack of education, some of the beneficiaries complained that they did not fully understand the legal content of the documents they signed at the NHA office. Furthermore, fatigue and hunger due to long queues under the heat of the sun in the provincial NHA office had deprived them the luxury of time to evaluate and understand the legal implications of the housing documents they signed.

Not comprehending the legal documents fully could deny the legitimate beneficiaries legal security to their shelters. Relocatees may possess stubs and signed documents, yet their names may not appear in the Alpha list (official list) of true housing beneficiaries of the NHA. In a 2002 Report of the government's Commission on Audit (COA), for instance, the audit team discovered that the NHA conducted no regular monitoring and evaluation of the actual occupants of the awarded units, since the latter believed that this responsibility belonged to the homeowners' association. Because of this, some families occupying the relocation were not on the NHA master list. If beneficiaries are not included in this list, double or multiple awarding of shelter units could not be detected, and unscrupulous activities perpetrated by housing syndicates or professional squatters could not be monitored and detected by authorities.

The officers of the Housing and People's Development Office (HPDO) of Rodriguez, Rizal, said that they have asked the NHA for the Alpha list of all the legitimate homeowners inside the Southville Rodriguez Housing Project (SRHP) relocation area. However, the NHA refused to accommodate their request. Thus, the HPDO suspected that the NHA must have been hiding something for refusing to provide the municipal government with the necessary data regarding actual and legitimate occupants of the SRHP. Due to the lack of information and profile of all relocatees allowed by the NHA to occupy the housing units, Barangay San Isidro initiated a census of all beneficiaries living in the entire relocation area in November 2012 (Fig. 6.1).

Professional squatters and squatting syndicates can exploit the NHA's lack of an updated and complete official list of legitimate homeowners and residents in the SRHP. They can occupy and buy housing units intended for disaster victims and

Fig. 6.1 Some of the unfinished and unoccupied housing units in Southville Rodriguez Housing Project (SRHP) Phase 8A. *Source* Author

later sell or lease them for a profit. Syndicates can also hire occupants masquerading as disaster victims to occupy the housing units in the relocation with different names appearing in the NHA's master list. One time, according to the Municipal Disaster Risk Reduction Management Office (MDRRMO) chairperson and head of the HPDO, the local police of Rodriguez were able to apprehend bogus relocatees and professional squatters whose names did not appear on the NHA master list. This incident confirmed some local officials' suspicion that big-time syndicates with connections in the NHA may have taken advantage of the government‘s low-cost housing and post-disaster recovery program.

This practice of double or multiple awarding of housing rights is common in the Philippine housing procedures wherein the real owners and occupants of the housing units are unaware that the units they amortisized from the government are actually titled in the name of another person. The possession of entry pass without the required legal documents is not what the law intended. Two key informants in SRHP revealed that they bought their units from the NHA after borrowing money from informal lenders. But they were shocked when they discovered that the houses they bought from the NHA local office were already owned by other people upon verifying the official list in the main NHA office.

6.2 Habitability

The right to an adequate housing by displaced persons such as disaster victims, as defined by the United Nations, does not only mean owning a house in resettlement areas but also acquiring a residential unit that is habitable and resilient against any disaster. Habitability implies that the housing has sufficient space and is structurally sound for protection against the destructive elements and disease vectors. The 2010 PDRRMA did not explicitly elaborate the type of houses that the developers must build in the relocation site. It assumed that all existing laws on housing construction under the National Building Code (NBC) and other real estate laws would be applied to build habitable houses for the relocatees. Habitability of dwelling units includes following standard building materials and procedures. The burden of monitoring whether housing units built by private developers in the SRHP comply with the building code and environmental requirements belongs to the NHA given the lack of authority of the LGU to interfere in the construction and management of the relocation site.

In the SRHP, Typhoon Ketsana victims complained that their houses have construction defects and are made of sub-standard construction materials. The house structure has no posts. Only two parallel walls, built with hollow blocks and a few small round steel bars partially filled with mixed cement, served as the "main posts" of the dwelling unit. To save money, the developer used only a few round steel bars around the house (8 mm). To save cement, the mixture of the sand, gravel, and cement did not follow the standard proportion, thus making the entire concrete structure weak, according to some construction workers. Moreover, the

Fig. 6.2 A sinking floor due to the absence of steel support and movement of the soil. *Source* Author

Fig. 6.3 A cracked bathroom wall caused by gradual movement of the ground. *Source* Author

cement floor did not have steel bars. Thus, some floors were sinking because some parts of the relocation were only reclaimed by the developer (Fig. 6.2).

According to some informants, many raw houses of the SRHP Phases 8A and 8B were built on top of reclaimed areas as the relocation area was formerly a farmland. The housing units were built on a weak soil foundation, and thus cracks are common inside the row houses. A report by Demolitionwatch (2012) on the condition of houses in the SRHP Phase 8B confirmed the weak construction of the houses. It reported that the housing units have cracked floors and walls, exposed sockets and fuse box, damaged door knobs, leaking water pipes, very thin galvanized roofing, and substandard steel support bars, i.e., just 8-mm hollow blocks 4 in. in width[1] (Fig. 6.3).

[1]See, "Urgent Appeal filed to the United Nations Special Rapporteur on Adequate Housing Regarding the Worsening Condition of Families in the Relocation." Available at http://www.demolitionwatch.

wordpress.com/ 2012/01/10/urgent-appeal-f-2/.

Fig. 6.4 A homeowner points to the cracked wall of his housing unit because of the gradual sinking of the ground. *Source* Author

According to some informants, the construction of housing units was *tinipid* (made very cheap) to earn more profit for the developers, contractors, and sub-contractors. Another major reason why the houses were very weak in materials and construction is that the developers allowed subcontractors to construct the housing units. To earn more money, the sub-contractor further lowered the construction cost by using sub-standard construction materials and disregarding the technical specifications and construction standards set by the National Building Code and NHA guidelines.[2] (Fig. 6.4).

The practice of using sub-standard materials in the construction of low-cost housing at relocation sites is often caused by the passing of contracts (sub-contracting) from one contractor to another. In the construction of infrastructure projects in the Philippines, it is a common practice of contractors to pass the work from one contractor to another and, in the process, a certain percentage of the project value is retained by each contractor and sub-contractors, thus resulting in the use of substandard materials or even unfinished projects.[3] In order to make more money, sub-contractors usually decrease the production cost by reducing the number of steel bars, amount of cement, size of hollow blocks, etc. This practice is a

[2]A technical team that inspected the houses in Kasiglahan Village, a relocation site also built by some developers of Southville Rodriguez, discovered that "the houses built for the communities did not have the adequate structural elements, which are required by the National Structural Code of the Philippines (NSCP), such as concrete reinforced columns needed to keep houses sturdy against earthquake and wind loads. Ground settlement can be seen everywhere. Furthermore, the spacing of the rebar reinforcements of the concrete hollow blocks did not match those required of the NSCP. Moreover, the increase of the row houses with respect to the nearby river is suspected to be inadequate for flood protection." (http://agham.org/982A135D-126E-4030-968A-8A7E39096724/FinalDownload/DownloadId-87F163CF5826B2B1D9575EF1D2454395/982A135D-126E-4030-968A-8A7E39096724/sites/default/files/agham-downloadables/kasiglahan_village_report_0.pdf)..

[3]http://www.pctc.gov.ph/papers/graft.htm.

win–win situation for the NHA's accredited developer–contractor of the relocation site. Developers are assured of payment from the NHA for all the units they built inside the relocation. By hiring sub-contractors, the developer earns profit without effort and gains capital to build more housing units. According to a conservative estimate of one sub-contractor in 2013, the actual price of the house and lot in the relocation was approximately PHP 40,000 (around US $800). But the selling price of the developer for each unit was between PHP 180,000–200,000. Whatever the agreement between the developer and sub-contractor, the developer still earned profit at the expense of the habitability of the houses for Typhoon Ketsana victims and other residents in the SRHP.

6.3 Affordability

The Affordability in low-cost and post-disaster housing requires that the housing units, including necessary services, be commensurate with the income level of the beneficiaries.

The payment schedule or monthly amortization of the houses in the SRHP as shown in Table 6.1 looks very affordable. However, if one considers the victims' income, number of children in the family, monthly expenses, and job instability, the monthly amortization scheme, which increases every 5 years, is not affordable for the beneficiaries. The daily fare to work, which costs PHP 60–150, eats up a large chunk of the worker's income. This is one main reason why some relocatees prefer to sleep in public places and return home only occasionally to save extra money for the amortization of their homes. The major problems for relocatees are unemployment and lack of stable jobs and livelihood. Even if the daily amortization may look very low, the average income of the relocatees, as indicated in Table 6.1, is really not enough for the daily expenses of their families. The government's Commission on Audit (COA) had already recommended to the NHA in 2004 to lower the amortization rate because this resulted in the accumulation of unawarded units and low collection. It also mentioned that the pricing was not affordable to the lowest 30% of the urban wage earners whom the NHA intended to serve.

Table 6.1 Southville Rodriguez Housing Project (SRHP) housing amortization

Year	Total monthly amortization (PHP)	Daily amortization (PHP)
First–5th	200	6.67
6th–10th	400	13.33
11th–15th	500	16.67
16th–20th	650	21.67
21th–25th	800	26.67
25th–30th	809.53	26.98
Total cost	201, 571.80	

Source NHA Kasiglahan Village I Office, Rodriguez, Rizal

Table 6.2 Typhoon victims' occupation and average daily income

Occupation	No. of workers	Average daily income (PHP)*
Construction worker	15	300
Driver	6	300
Carpenter	2	350
Electrician	1	420
Cook	3	200
Security Guard	1	400
Bus conductor	3	250

*Exchange Rate as of July 15, 2013: $1 USD = PHP 40.55

Except for the security guard, most of the jobs shown in Fig. 6.2 were casual, and their income only averaged around PHP 300, construction and carpentry work are paid per project and thus seasonal in nature. When a project is finished, the worker must find another job. If the worker becomes sick or incapacitated, the family experiences daily hunger as in the case of Aling Bettyline. Her husband had just been accepted by a construction firm but was dismissed by the employer immediately on learning that he had a heart ailment. He stayed at home depressed and contemplated suicide. There are no clinics, medical personnel, or free medications beyond first aid in the relocation area. The nearest public hospitals are 20–30 km away from SRHP.

6.4 Summary

This chapter has illustrated that the post-disaster housing for Typhoon Ketsana victims and relocatees had serious problems with regard to security of tenure, habitability, and affordability of the housing units. The beneficiaries were not given copies of the titles or contracts-to-sell by the NHA or developers. Only a piece of paper or stub served as the guarantee that relocatees have access to their houses. With the lack of an updated and comprehensive database of beneficiaries in the NHA, double awarding of units happened, and professional squatters and syndicates exploited this problem in order to deprive typhoon victims their right to ownership. Aside from security of tenure, the habitability of the housing units was also problematic. Developers, in order to increase their profit, used sub-standard construction materials and built incomplete and weak housing units in a reclaimed land, thus posing dangers to the lives of the occupants. They also hire sub-contractors to finish huge projects that further deteriorated the quality of the relocation houses. Finally, despite the low monthly amortization of the housing units, the housing project remained unaffordable to the relocatees. Displaced from their original jobs, and with low income in their new informal employment in the relocation area, relocatees were forced either to lease or sell their homes, thus defeating the original purpose of PDRRMA and UN provisions to provide affordable, safe, and resilient housing to disaster victims under the “build back better” principle.

Reference

Demolitionwatch. (2012). “Urgent Appeal filed to the united nations Special Rapporteur on Adequate Housing: Regarding the Worsening Condition of Families in the Relocation.” Retrieved from https://demolitionwatch.wordpress.com/2012/02/29/urgent-appeal-f-2/

Chapter 7
Jobs and Livelihood Program in the Resettlement Area

Abstract This chapter investigated the availability of jobs and livelihood in the government-owned Southville Rodriguez Housing Project (SRHP) area in the Philippines for Typhoon Ketsana victims. It started with the case of Bettyline as an illustration of the extreme poverty inside the relocation site due to the lack of employment opportunities and livelihood programs. Despite the assurance of the 2010 Philippine Disaster Risk Management Act (PDRRMA) and other official laws that disaster victims would be provided with a suitable location site and adequate housing with employment and livelihood opportunities under the "build back better" principle, poverty persisted in the SRHP. This chapter clarified the role of the various government agencies in providing jobs and livelihood to disaster victims in relocation sites owned by the government's National Housing Authority (NHA) but under the jurisdiction of the local government. Despite the plurality of official laws requiring livelihood programs for relocatees, the chapter's assessment showed that under the present set-up, it is ultimately the semi-autonomous social fields or the social connections between private developers and the government regulating bodies, such as the NHA, that determined the presence or absence of jobs and livelihood in the relocation area. The private developer, with the approval the NHA, had the power to decide whether or not the relocation site would be proximate to employment zones. Moreover, there are no explicit legal provisions in Terms of Reference (TOR) that required developers to comply with the employment and livelihood provisions of the law before building relocation houses and selling them to the NHA. The informal rule of *kanya-kanya* (individual or group initiative) sidestepped the official law and created unintended consequences to post-disaster recovery of the Typhoon Ketsana victims in the relocation area.

Keywords Post-disaster housing · Relocation · Livelihood · Legal pluralism · Filipino values · Jobs

V.O. Ballano, *Law, Normative Pluralism, and Post-Disaster Recovery*,
DOI 10.1007/978-981-10-5074-9_7

7.1 Case: The Story of Aling Bettyline and Family

The story of Aling Bettyline is one of the many stories of suffering, lack of jobs, and absence of sustainable livelihood program in government-owned relocation areas in the Philippines such as Southville Rodriguez Housing Project (SRHP) relocation area. Bettyline said she saw the author one morning conducting interviews in the site and she was hoping that afterward he might come back. Fortunately, he went back to the site that afternoon to follow-up his interviews with two key informants. Aling Bettyline said she had prayed and tried her luck to find him. She said that she had been waiting for him at the the entrance of the SRHP waiting for his help. She said that her family had nothing to eat for supper on that evening. She was worried for her husband and two of her six children who were sick. She was hoping that he could help her to buy food. Luckily, the author had some kilos of rice and canned goods in his car, which he usually gave as a token of gratitude to informants after interviews. He learned that her husband, the only breadwinner, who worked as a casual construction worker in the relocation area, was just terminated from employment because his employer discovered that he had a heart ailment. Her husband experienced difficulty in breathing. She said she could not bring him to a doctor for a check-up. The nearest public hospital from the SRHP was Amang Rodriguez Hospital in Marikina City, approximately 25 km. from the relocation site. They had no money for the transportation fare. In fact, they worried for their next meal, especially that her husband is now sick and unemployed. Said she was too shy to approach her neighbors for help because they are in serious financial trouble too. Besides, they had already helped her many times and lent her some money. Bettyline has no full-time or part-time job or any gainful livelihood since they transferred to the relocation site. She said she had participated in soap-making seminars and other skills training in the site. However, there was no capital or support from the government to start her own small business. Asked about their future in the relocation site, Aling Bettyline said that she just prayed and hoped that God would saved them from hunger and poverty (Figs. 7.1, 7.2 and 7.3):

> *Hindi ko po alam kung papaano pa kami mabubuhay dito sa relokasyon. Sa Diyos na lang kami umaasa na hindi niya kami pababayaan. Nag-aalala kami sa aming kalagayan sa ngayon sa pabahay. Napabayaan po kami ng gobyerno. Walang permanenteng trabaho, walang sapat na kita. Paano matutugunan ang aming pangangailangan lalo na sa mga batang nag-aaral pa, paano na ang kinabukasan nila?* [I don't know how we can continue to survive here in the relocation. God is our only our hope that He will not abandon us. We are worried about our current conditions in the housing. The government has abandoned us. No permanent job, no sufficient income. How can our needs be satisfied especially for our children who are still studying, what about their future?].

This story of Bettyline illustrates the serious problem of unemployment and lack of livelihood in government-owned relocation sites such as the SRHP. The official laws state that the employment opportunities and livelihood programs must be ready before the disaster victims can move into the relocation site. However, a closer look at the actual situation in the resettlement revealed otherwise. The promise of the

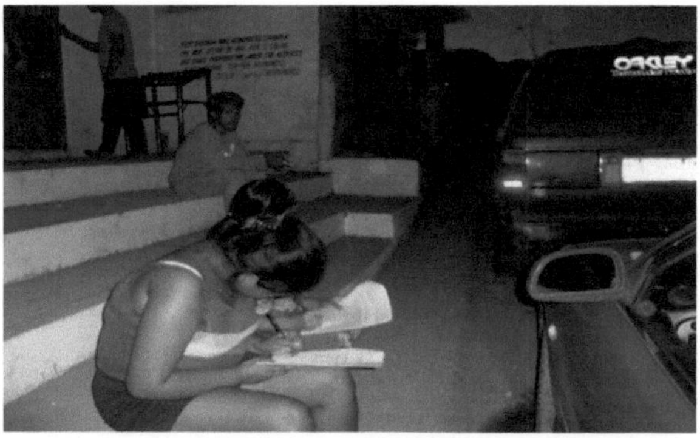

Fig. 7.1 Aling Bettyline, together with another Typhoon Ketsana relocatee, writing their personal story and experience concerning the lack of employment and livelihood in the SRHP in front of the San Isidro Chapel inside the relocation area. *Source* Author

Fig. 7.2 Aling Bettyline's housing unit in the SRHP Phase 8A. One can notice the darkness that surrounded the neighborhood due to the lack of electricity and street lights in the relocation area. *Source* Author

Fig. 7.3 Aling Bettyline's sickly child opening the main door of the house. This house was just bought two years old, and yet one can see the deterioration and cracks due to the poor quality of the building materials and construction defects. *Source* Author

Philippine Disaster Risk Reduction Management Act of 2010 (PDRRMA) and its affiliate laws on the post-disaster recovery of disaster victims remained elusive for Typhoon Ketsana victims in the SRHP.

7.2 Plurality of Laws on Employment and Livelihood in Relocation

A number of laws that deal with urban poverty and disaster state that urban poor relocated in government resettlement areas must have jobs and livelihood opportunities. However, due to legal pluralism, one would be confused concerning which law really governs the employment and livelihood programs of relocatees in the government resettlement site.

The PDRMMA contains only some broad provisions on the economic welfare of disaster victims after the disaster. However, its implementing rules issued by the National Disaster Coordinating Council (NDRMC), Section 2 (p) explicitly provides that the government must "implement emergency rehabilitation projects to lessen the impact of the disaster, and facilitate the resumption of normal social and economic activities." This provision implies that disaster victims must be provided with employment and livelihood opportunities in the relocation site. The Urban Housing and Development Act of 1992 (UDHA) also states that:

> Section 22. **Livelihood Component.** - To the extent feasible, socialized housing and resettlement projects shall be located near areas where employment opportunities are accessible. The government agencies dealing with the development of livelihood programs and grant of livelihood loans shall give priority to the beneficiaries of the Program.

An important housing-support for the resettlement program is livelihood assistance. Livelihood assistance is unique to the resettlement program and is justified in support of the displacement or dislocation of families. The livelihood expenditure includes the capital outlay for the construction of livelihood facilities usually consisting of a livelihood center, a tricycle, jeepney, or transport shed, and/or a *talipapa* (wet market) center.

In addition to livelihood infrastructure, the National Housing Authority (NHA) also allocated approximately PHP 3000/beneficiary household for capacity building or skills training. This budget is not included in the livelihood-item expenditure and is part of administration costs. For livelihood programs, the NHA acts as a facilitator, resource integrator, and planner. The NHA basically links the resettled communities to skills training, job placements, scholarship programs, and livelihood-based projects including credit or loan assistance of concerned agencies. This role is critical to enable the resettled communities to be mainstreamed into local and national programs. The NHA observed that the devolution of the programs of the Department of Trade and Industry and Department of Labor had created a gap in support and linkage at the micro or barangay level. In many cases, Local Government Units (LGUs) have focused their attention on the macro-investment aspect (i.e., attracting locators) without clear programs on the micro aspects; thus, the NHA is engaging the LGUs to ensure that resettled communities, in particular, are not left out. So far, the NHA has served or assisted a total of 63,000 beneficiaries of resettlements sites in Cavite, Laguna, and Rizal on skills training, job placements, livelihood programs, and grants. However, only 20% of households, at the most, in the mentioned resettlement sites are able to avail themselves of livelihood-enhancement programs at any one time. In addition, information about the use of the livelihood infrastructure is limited. Whether these livelihood interventions have resulted in employment or improved welfare cannot be determined from the information available and thus require further study (Ballesteros & Egana, 2012, p. 12).

The availability of jobs for economically disadvantaged disaster relocatees is a common problem confronting the government relocation program. The government usually prefers relocating disaster victims in remote provincial areas rather than in low-cost condominiums or tenements that are proximate to victims' employment or livelihood. Because most of the disaster relocatees are urban poor who live in squatter communities to provide cheap labor and services in the city's industrial or commercial districts, relocating them in distant places implies displacement and loss of jobs. It is therefore imperative for the government to provide alternative jobs and livelihood, even better than their previous source of income, for the victims in accordance with the "build back better" principle. However, what actually happened to the economic condition of the Typhoon Ketsana victims in post-disaster recovery is quite the opposite to what the law promised. In the SRHP, many victims experienced greater difficulty in finding stable jobs and an alternative livelihood. The estimated cost of fare is about 30% of the daily income if one is employed in the metropolis outside the relocation site. Thus, many male breadwinners choose to work as casual construction workers at a low provincial wage rate inside the relocation site.[1]

A visit to the NHA Website reveals that the focus of the NHA's program for livelihood consists mainly of skills training and often stops here. However, these livelihood seminars, informants said, were not realistic and sustainable. There is no long-term plan or sustainable livelihood program for the beneficiaries, especially for women, the necessary capital or loan and marketing support. Thus, they end up spending their hard-earned money on the training but have no opportunity to apply the skills they learned from the seminars. Other private organizations also help in sponsoring livelihood seminars in the area, but none of these initiatives have substantially improved the economic status of the residents. Many are still expecting the government to provide them with sustainable livelihood programs.

The various substantive laws and government agencies on socialized housing all speak about providing livelihood to the poor disaster victims in relocation sites.

[1]See Table 6.2 of Chapter 6 of this book.

However, in practice, what agency is really the overall in-charge of the livelihood program inside the SRHP relocation site?

7.3 Responsibility for Providing Employment and Livelihood

The perennial problem of the Philippine relocation program for the urban poor is the availability of livelihood in the resettlement area. The urban poor, which consists of the majority of disaster victims, are provided "dirty jobs" in the industrial and commercial centers of the urban economy. Resettlement in remote places means displacement from their existing livelihood. Thus, if they are relocated by the government away from their place of employment, the first thing the law must consider is the creation of alternative livelihood opportunities. The 2010 PDDRMA, in its definition of rehabilitation, briefly and indirectly provides that the government must initiate measures that ensure the ability of affected communities/areas to restore their normal level of functioning by rebuilding livelihood and damaged infrastructures and increasing the communities' organizational capacity (Section 3 [ee]). However, determining who leads in the rebuilding of livelihood for disaster victims and how can this be performed is unclear in the present law. Under normal conditions, usually the NHA has the organizational authority to provide livelihood to relocatees. However, during major natural disasters, conflict of authority arises between established organizations and outside and emergent groups (Quarantelli, 1986) (Fig. 7.4). Thus, the NHA, the LGU, and public and private organizations often compete for livelihood programs or blame one another for the lack of economic opportunities in the relocation areas.

The 2010 PDRRMA also mandated barangays and municipal disaster organizations to provide livelihood assistance to relocatees. In the case of Typhoon

Fig. 7.4 The Pledge of Commitment of the Housing and People's Development Office (HPDO) to provide security of tenure for its homeless and landless constituents through a comprehensive housing program and access to livelihood. *Source* Author

Ketsana relocatees, there seem to be discordant efforts between private organizations, the NHA, the Barangay Disaster Risk Management Office (BDRRMO), and the HPDO to extend livelihood to the relocatees. Some key informants revealed that these organizations occasionally sponsored livelihood training and seminars in the SRHP. Some revealed that they participated in soap-making seminars sponsored by the government in the relocation area. However, these livelihood seminars, they said, were not realistic and sustainable. There was no clear source of capital for raw materials and marketing plan. Thus, they ended up spending their hard-earned money on the training, but they had not started their own soap-making business. Other private organizations also helped in sponsoring livelihood seminars in the area, but none of these initiatives had apparently improved the economic status of the Typhoon Ketsana relocatees because all of them still expect the government to provide them with sustainable livelihood programs. Whose primary responsibility is it to develop and administer livelihood program for disaster relocatees in the SRHP?

In interviews, both barangay and HPDO officials admitted that they have some role to play in the rebuilding of livelihood of the disaster victims in the relocation area within their jurisdiction. However, because the SRHP is directly managed by the national government, they pointed to the NHA as the lead agency in the Typhoon Ketsana relocatees' livelihood program. The provincial office of the NHA in Kasiglahan has a committee whose main function is to provide training and livelihood seminars to the relocatees, not just from the SRHP but also to all neighboring relocation sites such as the Kasiglahan relocation areas. With thousands of relocatees needing livelihood assistance, funding became a major problem for the agency. This is also a major problem encountered by the BDRRMC and HPDO in providing sustainable livelihood to Ondoy and other disaster relocatees. Although the PDRRMA sets aside 5% of the total municipal or city government annual budget as a calamity fund, only 30% of this can be used as Quick Response Fund (QRF) for relief and recovery programs including livelihood (Section 21). According to key informants, this fund is insufficient for the growing number of relocatees in the SRHP, 90% of whom came from various parts of Metro Manila and neighboring towns. According to the HPDO head, this increased influx of outsiders has depleted the development and disaster funds of the town for disaster recovery and livelihood. These relocatees from other areas had a share in the Internal Revenue Allotment (IRA) of their respective local governments. However, citing legal and political considerations, mayors refused to share their disaster funds with the Local Government Unit (LGU) in charge of the relocation. As a result of lack of jobs and sustainable livelihood programs in the relocation area, some relocatees sold or leased their housing units in the SRHP and returned to their place of origin to resume their old informal work and trade, thus defeating the intention of the new disaster law to provide disaster victims with a better life in the relocation site.

7.4 *Kanya-Kanya* Attitude on Jobs and Livelihood

Section 22 of UDHA, under the livelihood component of the relocation program, explicitly provides that (1) socialized housing and resettlement projects shall be located near areas where employment opportunities are accessible and (2) government agencies dealing with the development of livelihood programs shall grant livelihood loans giving priority to the urban poor beneficiaries of the relocation. Section 23 further provides that the local government units in coordination with Presidential Commission for the Urban Poor and concerned agencies… shall encourage beneficiaries to organize themselves and undertake self-help cooperative housing and other livelihood activities in the resettlement area. However, in-depth interviews with relocatees in the SRHP and municipal officials of Rodriguez, Rizal, revealed that there was no clear entity in charge of the livelihood program of disaster victims inside the relocation site. According to the informants, there is no clear and sustainable livelihood program for disaster victims. Most of the livelihood programs initiated by private organizations consisted mainly of livelihood seminars, such as soap-making, but there is no marketing assistance nor loans for beneficiaries to start a business.

Although the UDHA is explicit that the local government unit, in coordination with the National Housing Authority, shall provide basic services and facilities and access to employment and livelihood opportunities to relocates that are sufficient to meet the basic needs of the affected families, there is still ambiguity of who is in charge of the livelihood program of a complex relocation system such as the SRHP. Neither the PDRRMA nor the UDHA is very specific on who exactly is responsible for the relocation and livelihood program of all disaster victims including the non-resident or even the "professional squatters" of a national resettlement site located in a local government unit as the Municipality of Rodriguez: Is it the NHA, or is the local municipal housing agency, i.e., the HPDO? This ambiguity leaves more discretion for public officers to interpret the law, which sometimes leads to conflict and indifference to the plight of disaster victims. Based on interviews, it appeared that the responsibility of providing livelihood programs inside the relocation site would depend on the origin of disaster victims and the initiatives of the public officers to produce funds for this purpose. Because of the tendency of local mayors to engage in *kanya-kanyang pamamalakad* (individual initiative) with regard to securing relocation and livelihood for their own constituents, there is no clear, direct, and overall supervisor (although the NHA is the administrator of the relocation area) entity that is in charge of managing and harmonizing various efforts to provide livelihood programs to the urban poor and disaster victims of the SRHP.

In the Municipality of Rodriguez, the local government, headed by the mayor, and the municipality's Housing and People's Development Office (HPDO) are directly responsible for the long-term recovery program for disaster victims in terms of socialized housing and livelihood programs. However, according to a Commission on Audit (COA) report this all depends, on the available budget, which in turn depends on the determination and creativity of the mayor in increasing funds beyond the 5% disaster fund from the annual municipal budget for disaster purposes. During the incumbency of a former mayor, for instance, proceeds from the town's dump site were used to build a relocation site for local informal settlers, and the Rodriguez HPDO was able to provide livelihood programs for the beneficiaries [39]. According to the provision of the Local Government Code on the autonomy of local government units (LGUs), the HPDO and the mayor exercise discretion, monitoring, and supervision of the relocation area within their jurisdiction for the attainment of their post-disaster recovery. However, for the relocatees who came from the cities or municipalities, the local government of Rodriguez does not provide relocation and a livelihood program. The informal policy of *kanya-kanya* prevails in providing relocation and livelihood to the SRHP. In this sense, the local government, which has the direct administrative power of the municipality, is no longer in total control the outcome of the post-disaster recovery program in its own proverbial backyard (Fig. 7.5).

The overlapping of laws and the lack of provision in the PDRRMA creating a super-agency that oversees and harmonizes the coordination of local and national government agencies in relocating disaster victims, as well as providing them with livelihood opportunities, is a major contributing factor of this prevailing *kanya-kanya* attitude of program implementers. Both short- and long-term post-disaster recovery programs require coordination. Thus, the local government of Rodriguez is faced with coordination and jurisdictional problems in governing a relocation site managed by the national corporation such as the NHA. Yet relocatees and migrants are technically under the administrative jurisdiction of the local government of Rodriguez once they resettle in the SRHP. Although the new PDRRMA explicitly intended that poor homeless disaster victims' should be given socialized housing and opportunities to lead a better life in a relocation area in accordance with the "build better" principle, the final outcome and implementation of this depends on the origin of the relocates, the discretion of local executives, and the availability of funds.

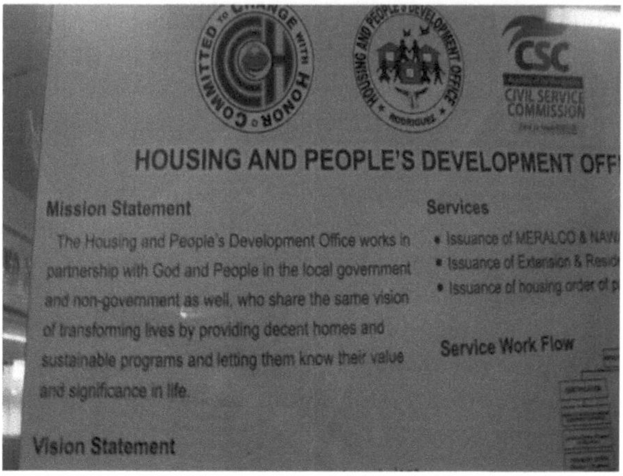

Fig. 7.5 Mission statement of the Housing and People's Development Office (HPDO) of Rodriguez, Rizal, promising partnership with local government and non-government organizations to transform lives by providing decent homes and sustainable programs. *Source* Author

Lastly, many local mayors are reluctant to lose political control over their constituents for election purposes, and some unscrupulous NHA officials are resistant to giving up their illegal practices (*kalakaran*) of bringing professional squatters to relocation areas. This situation reinforces the *kanya-kanya* attitude of various groups and agencies involved in the relocation system and results in conflict and difficulty of creating and implementing a comprehensive and uniform post-disaster recovery program for all disaster victims in the SRHP.

7.4.1 Availability of Jobs and Livelihood in the Relocation Area

Building resilient communities ready to recover after the disaster as envisioned by law requires that the relocation site must also be equipped with opportunities for employment or other economic activities as well as with health care, transport, and other social facilities. In Section 2 (p) of its implementing rules, the PDRRMA explicitly aims to… "implement emergency rehabilitation projects to lessen the impact of the disaster and facilitate the resumption of normal social and economic activities." This implies providing disaster victims with employment and livelihood opportunities in the relocation site. Section 2 (p) of the implementing rules of the PDRRMA explicitly provides that that there must be emergency rehabilitation projects to lessen the impact of the disaster and facilitate the resumption of normal social and economic activities. This implies that the government must provide disaster victims with employment and livelihood opportunities in the relocation site.

However, the availability of jobs for poor disaster relocatees is a common problem confronting the government relocation program. The government usually prefers relocating disaster victims in mass housing in remote provincial areas rather than in low-cost condominiums or tenements proximate to victims' employment or livelihood. Because of the disaster, relocatees are the urban poor who live in squatter communities and provide cheap labor and services in the city's industrial or commercial districts; relocating them in distant places implies displacement and the loss of jobs. It is therefore imperative for the government to provide alternative jobs and livelihood, even better than their previous source of income, for the victims in accordance with the "build back better" principle of the law. However, what happened to the economic condition of the Ketsana victims in post-disaster recovery is quite the opposite to what the law promised. In the SRHP, many victims experienced greater difficulty in finding stable jobs and alternative livelihood after the disaster. "Some retain their work and source of livelihood in Metro Manila but spend very high on transportation ranging from PHP 70 to 250 going to work and back to the relocation site. Others rented rooms near their work and spend PHP 500–1000 per month" (Demolitionwatch, 2012). Thus, many male breadwinners chose to work as casual construction workers at a low provincial wage rate earning

Fig. 7.6 Small *sari-sari* (variety) stores at the entrance road of the SRHP. *Source* Author

PHP 150–200 a day with no social benefits. The women in the relocation site strived to live by setting up a *sari-sari* (variety) store and selling food inside the community. However, many shops have already closed due to lack of capital and unpaid debts (Fig. 7.6).

7.5 Summary

This chapter has shown that despite the Philippines' holistic approach to disaster management and legal provisions of the PDRRMA, the UDHA, and other related laws that promised jobs and economic assistance to urban poor and disaster victims, permanent employment and sustainable livelihood remained elusive to the Typhoon Ketsana victims in the SRHP. The task of providing a sustainable livelihood to the beneficiaries became a major source of conflict between national and local government agencies tasked to provide a long-term post-disaster recovery to disaster victims. Owing to overlapping of laws and disputes on jurisdiction, Typhoon Ketsana victims still did not have permanent jobs and profitable alternative livelihood to improve their lives. Jobs available for the relocatees were mostly casual and low-paying. Many of the livelihood programs given by public and private agencies to relocatees consisted mainly of livelihood training and seminars without clear funding and marketing assistance. Some relocatees opted to return to their former jobs and livelihood activities in Metro Manila and nearby cities, thus resulting in the sale and leasing of their housing units in the relocation area.

References

Ballesteros, M. & Egana, J. (2012). Efficiency and effectiveness review of the National Housing Authority (NHA) resettlement program. Retrieved from http://www.dbm.gov.ph/62C97C27-20B5-4CB0-9BD2-8286EDF12907/FinalDownload/DownloadId-802ACD5E5502BED%20A5F66A855713/62C97C27-20B5-4CB0-9BD2-8286EDF12907/wp-content/uploads/%DBM%20Publications/FPB/ZBB-2012/e.pdf. Accessed March 4, 2015.

Demolitionwatch. (2012). *Urgent appeal filed to the United Nations special rapporteur on adequate housing: Regarding the worsening condition of families in the relocation*. Retrieved from https://demolitionwatch.wordpress.com/2012/02/29/urgent-appeal-f-2/

Quarantelli, E. L. (1986). Disaster crisis management. Preliminary papers #113. Disaster Research Center, University of Delaware. Retrieved from http://udspace.udel.edu/bitstream/handle/19716/487/PP113.pdf

United Nations (n.d.). *Basic principles and guidelines on development-based evictions and displacement*. Retrieved from http://www.ohchr.org/Documents/Issues/Housing/Guidelines_en.pdf

Chapter 8
Adequacy of Social Facilities and Services

Abstract This chapter examined the adequacy of the social facilities and services of the government-owned Southville Rodriguez Housing Project (SRHP) relocation site for some Typhoon Ketsana victims in the Philippines. In particular, it evaluated the availability and sufficiency of the relocation's electricity, water supply, health services, as well as educational, hospital, and religious facilities, according to the post-disaster recovery and "build back better" (BBB) principles promised by the Philippine Disaster Risk Reduction Management Act of 2010 (PDRRMA) and Urban Housing and Development Act of 1992 (UDHA) for the relocation of internally-displaced persons. It also clarified the role of the National Housing Authority (NHA), Local Government Units (LGUs), and government-accredited developers of resettlement areas in providing these social facilities and services. It took notice of the unintended effects of the plurality of laws and norms to the coordination of public and private agencies tasked to provide adequate housing to typhoon victims in resettlement areas.

Keywords Adequate housing · Relocation · Social facilities · Post-disaster · Social services · Typhoon Ketsana

8.1 Introduction

The Philippine Disaster Risk Reduction Management Act of 2010 (PDRRMA), following the international conventions of United Nations Commission on Human Rights, stipulate that there is no adequate housing for disaster victims if there is no availability and adequacy of services, materials, facilities, and infrastructure in the relocation area that satisfies the socio-cultural norms of the disaster victims. This implies that occupants must have access to safe and drinking water, adequate sanitation, and energy for cooking, heating, lighting, food storage, and/or refuse disposal. The right to adequate housing does not just mean adequate structure of the house itself. It includes sustainable and non-discriminatory access to facilities essential for health, security, comfort, and nutrition. And this means access to safe

V.O. Ballano, *Law, Normative Pluralism, and Post-Disaster Recovery*,
DOI 10.1007/978-981-10-5074-9_8

drinking water, energy for cooking, heating, lighting, sanitation, and washing facilities, means of storing food, refuse disposal, site drainage, and emergency services (OHCHR, 2009, pp. 8–9). This chapter examined the socio-cultural program in the Southville Rodriguez Housing Project (SRHP) in Rodriguez, Rizal and explained how the many rules and informal cultural norms, such as *palakasan* (patronage), influenced the implementation of the PDRRMA and other related laws on adequate social facilities and services for disaster victims in resettlement areas.

8.2 Laws on Social Services in Relocation Areas

Under Article V, Sections 21 and 22, of the Urban Development and Housing Act of 1992, the NHA must "[p]rovide basic services and facilities and access to livelihood opportunities sufficient to meet the basic needs of the affected families in socialized housing or resettlement areas in cooperation with private developers and concerned agencies." However, this responsibility of providing basic services and facilities is not the sole responsibility of the NHA. The law says that this will be performed by the NHA "in cooperation with private developers and concerned agencies."

However, the contractual relations between private developers of relocation areas and the NHA in their Terms of Reference (TOR) did not include a provision that requires the developer to ensure that the relocation site is equipped with adequate basic services and facilities before the beneficiaries or disaster victims move into their housing units in the resettlement. In the absence of this specific provision, which included penalties or even cancellation of the contract, the NHA could not coerce the developer to provide basic services and facilities in the relocation area.

The Local Government Unit (LGU) is also tasked by their own local laws to provide basic services to their constituents in the relocation areas. The municipal government of Rodriguez, Rizal, through its Housing and people's Development Office (HPDO), is also mandated to provide basic services and livelihood to their local relocatees. The absence of a unified disaster law on relocation and housing that harmonized other local laws has created loopholes in the recovery program: What public agency is really responsible of providing adequate social services and facilities to disaster victims in Philippine relocation sites? In case the NHA, LGU, or the developer neglected their tasks of giving social services in the relocation area, who will takeover this responsibility for the disaster victims? In the absence of a one-stop government agency on post-disaster reconstruction that supervises the entire rehabilitation process to ensure that disaster relocatees receive adequate social services, there is no assurance that the legal standards of the PDRRMA and disaster laws on social services as part of the right calamity victims to adequate housing can be met.

8.3 The Social Services Situation in Southville Rodriguez

A study by the Center on Housing Rights and Evictions (COHRE), an international human-rights organization based in Geneva, Switzerland, revealed that the living conditions in relocation sites in the Philippines are appalling due to lack of basic services such as potable water, electricity, and sanitation facilities (COHRE, 2008). This is confirmed by disaster victims in Southville Rodriguez (SRHP). Typhoon Ketsana victims in the SRHP complained that the pipe water and electricity were only available after two years of living in total darkness in the resettlement area, using only candles and kerosene lamps to illuminate their homes that further increased their daily expenses.

Under the present Philippine resettlement set-up, the NHA undertakes housing development, whereas estate management/building maintenance of the relocation area shall be undertaken either by LGUs, NGOs, or homeowners associations. Although the responsibility of housing development, as well as providing the necessary infrastructure for social facilities and service, primarily lies on the shoulders of the private developer who designed and constructed the relocation site and housing units, there is still a gray area regarding who will exactly assume the responsibility in case the NHA, developers, LGU, and other agencies failed to perform their duties as stipulated by law. With a plurality of formal and informal rules, it is difficult to blame a specific agency or group for any negligence done to the relocation and housing project. Like employment and livelihood, the Terms of Reference (TOR) between the NHA and the private developer does not stipulate the responsibility to provide adequate social facilities. The TOR only stipulates that the developer conforms with the government's environmental standards in building the relocation site and housing units but not concerning social facilities and services.

The duty of providing electricity, lighting, water—especially potable water—for site drainage, washing facilities, and other services and facilities in the relocation site belongs to the developer. In the case of SRHP, especially Phase 8A and 8B, this duty belongs to its two NHA-accredited developers: San Jose Builders and Baque Corporations. These facilities are actually part of the relocation facilities as stipulated in the contract between the NHA and the developers. Under the law, there is no complete delivery of the housing project if relocation facilities are absent or incomplete. Yet the NHA and developers allowed the Typhoon Ketsana victims to move in to the SRHP in 2011 even without electricity, water, and other social facilities in the relocation site.

Because the housing and relocation projects of the accredited developers are under the control of the NHA, the latter could pressure and demand that the former fully provide the necessary social infrastructure and facilities in the relocation site before the disaster victims and other beneficiaries occupy their housing units. Otherwise, the housing contract can be terminated by the government and NHA because of the breach. With regard to health facilities and services, as well as peace and order in the relocation site, the responsibility belongs to the barangay and the municipal governments.

There is an ongoing conflict between the LGU of Rodriguez and the NHA. According to barangay and municipal officials, the NHA does not cooperate with them and provide them with an Alpha list or official list of all people relocating in the SRHP. They surmised there might an unscrupulous power elite inside the NHA who are illegally connected with some housing syndicates who want the LGU to be in limbo with regard to accurate data and statistics of people occupying the housing units in the relocation area to pursue their illegal sale of relocation houses.

8.3.1 Electricity

Almost all low-cost housing and resettlement projects of the Philippine government promise the availability of water and electricity before beneficiaries can transfer to the relocation site. However, this promise is often inadequately fulfilled. In reality, the beneficiaries usually wait for more than one or two years before they can have water and enjoy electricity. As already mentioned above, the Typhoon Ketsana victims in the SRHP only acquired electricity in the relocation site after almost two years of stay in the area. There was no electricity when they were transferred to the site. Because the relocation is a remote place and a former farmland, the area is indeed dark at night, especially with the narrow streets of the subdivision with no street lights. Moreover, many residents complained that the developer had made electricity in the relocation a profitable business. Residents do not have individual electric metering. The developers would just give them an unofficial receipts showing their high monthly electric bills. The electricity consumption of the residents due to common metering is often the main issue in local protest rallies of beneficiaries and some members of NGOs. Residents claimed that the unpaid bills of other homeowners were added to their monthly bill as a system loss, thus making electricity further unaffordable given their meager income.[1] The sub-meters are controlled by the developer. An individual monthly bill can reach as high as PHP 2000 (US$40) because of the additional system-loss charge. Without prompt payment, their electric connection is immediately cut off by the developer, according to some informants. Moreover, residents claimed that the developer would charge them a minimum of 20 kilowatt-hour per month and household, approximately equivalent to PHP 500 (US$10), i.e., greater than their daily wage or income. Every resident is required to pay this minimum charge even if their consumption is lower than 20 kilowatt-hour. Some of the residents work and stay in Metro Manila to work or engage in sidewalk vending during weekdays and return only to the relocation during weekends. Thus, they consumed electricity less than the minimum charge.

[1]Newsreel: Residents protest lack of electricity in the SRHP Phase 8B relocation site. See https://www.youtube.com/watch?v=kXu1q-i-TF0.

Fig. 8.1 The SRHP Phase 8A by day showing Manila Electric Company (MERALCO's) electrical posts with no street lamps. *Source* Author

Fig. 8.2 Southville Rodriguez Housing Project (SRHP) Phase 8A by night, showing the darkness of the relocation site due to the absence of street lights. *Source* Author

Because there is no clear proof of monthly electric consumption, the Typhoon Ketsana victims in the SRHP believed that they are overcharged by the developer in their electricity bills. In an investigation, Demolitionwatch group discovered that the developers Baque Corporation and New San Jose Builders of the SRHP used sub-meters instead of direct electricity connections. In the SRHP Phase 8B, for instance, the residents were seriously facing problems of irregularities, high payment of system loss, overpricing in the electricity bill ranging from PHP 16.75 to PHP 23 per kilowatt per hour and the compulsory payment of a flat rate per month for electricity (Demolitionwatch, 2012) (Figs. 8.1 and 8.2).

Fig. 8.3 One of the few artisian wells used by the SRHP Phase 8A residents before water facilities were installed in the area. *Source* Author

8.3.2 Water Facility

Like electricity, water facility was only available after more than one or two years after the Typhoon Ketsana victims were already relocated in the SRHP Phase 8A. The source of water is not the water concessionaire, Manila Water. However, a few artisan wells were available in the area when the relocatees transferred to SRHP Phase 8A. For the relocatees, this is a great source of inconvenience and suffering in the relocation. Because these pumps were meant to be used by only a few instead of the thousands of relocatees in Phase 8A, long queues were a common sight daily in the area. According to informants, some residents would wake up at dawn just to avoid the long queues at the artisan wells lines and to fetch water for their cooking, washing, and bathing of their children before attending school in the morning. The water is not safe and potable. Residents need to boil it, which implies extra expenses for charcoal and gas for each household. Cases of diarrhea and other intestinal diseases, especially among children, were common due to the poor quality of the water. During the first two years of their stay in the SRHP, residents get water from the artesian well or through water delivery from the developer. Drinking water is not, however, available to relocatees, contrary to the promise of the developer and the NHA that adequate social facilities will be provided before the disaster victims could occupy their housing units. The Typhoon Ketsana victims in the SRHP had no choice but to buy mineral or purified water at an average PHP20/day—another added expense—for drinking, especially for their children (Fig. 8.3).

8.3.3 Health and Hospital Facilities

Health services is another major problem in SRHP for its estimated 40,000 residents in 2013. The entire relocation area had no public and private hospitals and adequate clinics inside the relocation site nor in the nearby barangay center. According to informants, the barangay health center could only offer first-aid intervention. There were no permanent doctors or nurses stationed in the barangay health center for emergency cases. Medicines, aside from a few paracetamols, were not readily available to relocatees. There is no ambulance on standby for emergency situations. The nearest private clinic is located in Kasiglahan Village I, a few kilometers away from the SRHP, and the nearest public hospitals, which offer cheaper medical services, are the Amang Rodriguez General Hospital in Marikina City and the East Avenue Medical Center in Quezon City. These hospitals are located far from the SRHP, approximately around 15–20 kilometers away from the relocation site, and require at least one tricycle and two jeepneys or bus rides to reach. The high fare is a major reason why many poor relocatees only visit these government hospitals during emergencies or life-threatening cases despite free or low medical fees.

8.3.4 Educational and Religious Facilities

Relocationing disaster victims into the SRHP is often done in the middle of the school year. Many Typhoon Ketsana victims in Phase 8A were relocated sometime in December 2009 and January 2010 when their children had not yet finished the school year in their place of origin. Although there is an elementary and a high school in the relocation site, these schools were not prepared to accept transferees from the relocatees. The school buildings lacked chairs, tables, and classrooms to accommodate transferees. A fact-finding mission by Gabriela Women's Party revealed that parents forged agreements with teachers that their children would attend their classes school only twice or thrice a week or only during examinations just to finish the school year while minimizing daily expenses (COHRE, 2008).

Finally, the current PDRRMA, which claims to be holistic in its approach failed to include religious structures and facilities as an integral part of long-term post-disaster recovery of the beneficiaries and Typhoon Ketsana victims. The construction of religious churches or chapels to address the religious needs of the relocatees is not an integral part of the relocation plan. However, the NHA has the discretion of whether or not to allow the construction of churches inside the relocation area depending on the requesting party. According to one pastor of a small born-again Christian group in the SRHP Phase 8A, she requested that her group be allowed to buy a piece of land and build their church. But she said that the NHA had repeatedly rejected her request. But she heard some news that the powerful Iglesia ni Cristo (INC) Church was able to buy a large parcel of land in the SRHP Phase 8B for the construction of their new church (Fig. 8.4). Some informants confirmed that *palakasan* (patronage) plays an important informal norm in getting concessions and favors from the NHA. Those who are *malakas* (politically strong) or who have social and political connections with some top officers in the NHA can often get what they to do in the SRHP.

Fig. 8.4 Due to extreme poverty, many residents cannot afford to buy a liquefied-gas stove. Instead they use charcoal for cooking, which is even more expensive because wood charcoal is scarce. *Source* Author

Fig. 8.5 The San Jose Labrador Chapel of the Catholic Church was the only religious structure found inside the relocation in early 2013. However, this building already existed before the SRHP was constructed as a relocation site in 2009. *Source* Author

8.4 Summary

Despite the legal provisions of PDRRMA and its allied laws that assured social and educational facilities and services to disaster victims, Typhoon Ketsana victims experienced hardships inside the SRHP due to the lack of water supply, electricity, and health facilities in the relocation area. Adequate relocation and housing for disaster victims must not only provide housing units but also sufficient social, health, and educational services. This chapter had shown that relocatees and disaster victims did not receive adequate social services in the SRHP especially when they first moved into the resettlement area. The relocation site had no electricity or adequate water supply during the first and second year of relocatees' residence in the area. The health center in the area had no nurse or doctor. It can only provide first aid assistance, but no medicines. There are no public and private hospitals nearby. The relocation had no adequate religious structures to provide religious services to relocatees with different religious affiliations.

References

COHRE. (2008). Written comments by the Centre on Housing Rights and Evictions (COHRE) concerning Philippines for consideration by the United Cultural Rights at its 41st Session, 3–21 November 2008. http://www2.ohchr.org/english/bodies/cescr/docs/info-ngos/COHRE Philippines.pdf

OHCHR. (2009). The right to adequate housing. Fact sheet No. 21 (Rev. 1). http://www.ohchr.org/Documents/Publications/FS21_rev_1_Housing_en.pdf. Accessed February 12 2015.

Demolitionwatch. (2012). Urgent appeal filed to the United Nations special rapporteur on adequate housing regarding the worsening condition of families in the relocation. https://demolitionwatch.wordpress.com/2012/02/29/urgent-appeal-f-2/

MIX
Papier aus verantwortungsvollen Quellen
Paper from responsible sources
FSC® C105338

If you have any concerns about our products,
you can contact us on
ProductSafety@springernature.com

In case Publisher is established outside the EU,
the EU authorized representative is:
Springer Nature Customer Service Center GmbH
Europaplatz 3, 69115 Heidelberg, Germany

Printed by Libri Plureos GmbH
in Hamburg, Germany